the joy of
hobby
farming

the joy of
hobby
farming

GROW FOOD, RAISE ANIMALS, AND ENJOY A SUSTAINABLE LIFE

Michael and Audrey Levatino

Skyhorse Publishing

Skyhorse Publishing books may be purchased in bulk at special discounts for
sales promotion, corporate gifts, fund-raising, or educational purposes. Special
editions can also be created to specifications. For details, contact the Special Sales
Department, Skyhorse Publishing, 307 West 36th Street, 11th Floor, New York, NY
10018 or info@skyhorsepublishing.com.

www.skyhorsepublishing.com

10 9 8 7 6 5 4 3 2 1

Library of Congress Cataloging-in-Publication Data is available on file.
ISBN: 978-1-61608-228-4

Printed in China

For Tippy and Oswald

contents

Preface   x

Place

1. INTRODUCTION   3
Why Hobby Farming?—Slowly Farming

2. THE SEARCH FOR A FARM   9
Renting or Leasing—Location—Realtors and Other Resources—How to Look—Finding the Silver Lining—Evaluating a Farm's Water Profile—Zoning, Property Rights, and Easements—Taxes—Insurance

3. YOUR FARM   19
Where to Get Information—Wells, Springs, and Cisterns—Septic System—Generators—Heating with Stoves—Farm Safety—Guns and Hunting—Farm Equipment and Tools—Fencing—Gates and Cattle Guards—Outbuildings and Barns—Firewood—Ponds

(2)

Growing Things

(3)

The Care of Living Creatures

Running Your Farm as a Business

Preface

Imagine any sunny summer afternoon. Chances are, my wife Audrey and I have spent the day at the farmer's market, selling our selection of cut flowers, vegetables, jewelry, eggs, and honey. We wake up early, before the sun is up, to load the car under the lights of our shed. At the market, we set up our canopy and create our multi-level display on tables and wooden crates covered with tablecloths. We hang our farm banner, write the day's offerings on our white board, and design some flower bouquets in vases as we wait for the customers. The hours fly by as we sell our wares and there's rarely a moment to sit and rest.

After noon, when the market is shut down and we count our moderate sales, we again pack up the car with a (hopefully) much lighter load. Back at the farm, we spend an hour or so cleaning out our flower buckets, storing the leftover flowers in the cooler, and putting away our supplies. The profit for our time doesn't quite add up, but with my "real" job kicking back in on Monday morning, we can certainly justify a pleasant nap. We've learned that when it comes to hobby farming, we need to use the term "profits" loosely—the pleasures of an afternoon enjoying the land or our friends at the farmer's market easily evens out the costs of the work.

We weren't looking for a farm when we decided to move to Virginia. We'd been living in the Bay Area, where we'd relocated from Colorado during the height of the Internet boom years. We moved from the peace of the mountains to a frenetic, pulsing area of new wealth and utopian values. We both had new jobs, mine as a travelling book salesman and Audrey as a high school teacher in Silicon Valley.

We were young and we worked hard during those three years in Oakland. The traffic was hellish and Audrey commuted to work two hours each way. Every day was different for me as I fought the traffic to bookstores all over Northern California up to Washington State, sometimes being away from home for a couple of weeks at a time. But we settled into our one-bedroom apartment on the top floor of an art deco house. We immersed ourselves in the local food and music scene. We took on the local values by recycling (those were the days when not everyone did), forgoing meat and supporting animal rights. We became Bay Area denizens.

But we longed for our own home and were tired of paying a king's ransom for the rent on our tiny apartment. So when a sales territory at my company opened up in the mid-Atlantic region of the country, we jumped at it. And the search began. We wanted a starter home with at least a couple of acres of land for a garden and a dog.

After many weeks of searching, our farm found *us*: "Bring your animals. Two-bedroom house in the country on twenty-three acres. Fenced for horses." We'd never dreamed of owning a farm. But we suddenly realized that we now had the opportunity to put all our ideals into practice. We'd been eating local foods, but not growing them. We'd cared for one cat; now we could adopt many more animals. We'd donated to environmental organizations to protect land from development; now we had a golden opportunity to protect 23 acres of our own. We could become responsible producers instead of just responsible consumers.

As you'll read in these pages, we went for it. But we couldn't have done it without an off-farm job. I travel to the city once a month to check in with my company, doing the rest of my work as an account manager for a large publishing house from home or the road. Without my off-farm income, there would be no farm to call our own. And while the farm provides the cut flowers, vegetables, eggs, and honey we produce, it's also a healthy outlet at the end of a hectic sales week. It stretches muscles, both physical and mental, that become cramped from office work and travel. It's a lifestyle that provides balance.

PART ① {.part-marker}

Place

The view of our farm
in fall from the other
side of the pond.

CHAPTER 1

Introduction

The best piece of advice we got when embarking on our hobby farm was, "Start small and don't overwhelm yourself." There are countless stories of folks who move to the country, buy a flock of sheep and several horses, order a bunch of chickens, and start an organic garden, only to exhaust their economic and physical resources. It's much easier to start small and grow into your comfort level than to go "all in" and try to keep up. This is the key benefit of hobby farming. Since you aren't pressured to make it profitable right away, you have room to explore and grow into your farm.

Why Hobby Farming?

To *hobby farm* is to enjoy the bounty of your land without making economic demands on it (or yourself) that would degrade its natural sustainability. The ethos of hobby farming is living close to the land, protecting it from development and overproduction, savoring the bounty of what can be produced on it by your hands alone, and sharing the fruits of your labors with nearby friends and strangers.

Contributing to your local economy is also important. It's now accepted wisdom that our country is moving away from factory-farmed, chemical-laden foods and more toward locally grown, organic foods. If there's to be diversity at all in the offerings at your local farmer's market, a necessity if local farmers are to compete with the national grocery stores, then it will take a small army of hobby farmers offering a wide diversity of products. You can be one of these farmers, contributing bounty to your local food economy while enjoying your passion to work on your land.

Hobby farming embraces the idea that smaller is better. Better tasting foods, both plant and animal, come from small farms that don't use intensive cultivation methods to increase profits. Hobby farming profits come mostly from the reddest heirloom tomatoes, grown with personal attention and careful handling; the freshest eggs with firm, orange yolks, made by chickens who have fresh air and

room to roam; and the hardiest, most gorgeous flowers that retain their brilliance for weeks in a vase. The profits aren't always monetary—you'll have a stronger body and mind gained from hours of personal fulfillment working on the land.

Unless you're independently wealthy or have inherited your farm, you're probably going to keep or find a job outside the farm to pay your mortgage or rent. That's why it's called *hobby* farming. It doesn't mean you're not taking it seriously; it means that you're realistic and practical. After all, diving headlong into this particular economic situation (surviving on farming alone) has ruined countless well-intentioned people.

Of the hundred or so vendors at our thriving farmer's market (the largest in Virginia), only a handful are able to make a living solely by farming. Most profitable farms have been developed over generations. Families settled and began working in mostly poor conditions. They improved their land, invested in equipment, and passed a slightly more profitable farm with less expenses on to the next generation. But this farm improvement curve was broken by the factory farming model and many family farms have been lost. In addition, young people aren't farming, and there's no one to take over for the aging farmer population. And why should there be when the economics of farming don't work for anyone that has to pay rent or a mortgage? Farming is like any profession, and there needs to be a vibrant apprentice and training system that doesn't require investing your life savings for those that are considering farming as a profession.

To us and others like us, hobby farming is the key to reclaiming a local farm economy. Hobby farming offers a hands-on way to explore the skills and economics of farming while saving for that future. But we're typically starting from square one, with spent land and little equipment. This is why an off-farm job is so important; a farm needs to be retooled and its sustainability revitalized before it can again become an economically viable enterprise.

Hobby farming can keep rural land from becoming tract houses and suburban lots. These farmers keep heritage species alive, whether a unique and pure strain of squash or a chicken whose roots are centuries old. By farming the land, they protect countless species of animals and plants that already exist there, and they preserve skills and knowledge that would certainly fade into the hum of our rapid technological advances.

Unlike industrial farmers, who must deal in large economies of scale in cultivating a single crop (corn, soybeans, or cattle), hobby farmers dabble in a little bit of everything, like keeping chickens, horses, gardens, bees, mushrooms, timber stands, and fish ponds.

Anyone can be a farmer, from the college student growing and selling vegetables in the summer for tuition, to the retirees that can't stand another game of cards by the pool, to the professional exploring any and all possibilities in life before settling for any one of them, to the young family looking to raise kids to be self-sufficient.

Slowly Farming

Living and operating a farm is only as romantic or as hard as you want to make it. We started with just our two chickens, Ted and Bev. It took us months to get a dog. Then we adopted another. Soon we started building our vegetable garden and fenced it in using our own cut cedar trees as posts and borders for raised beds and the side of a packing crate for a gate. After hatching a clutch of eggs, we built a coop and grew our chicken population to eleven. Our field of grass grew so tall that we finally bought a used tractor after a full year on the farm. A freshly cut field begs for some animals to graze it, so we acquired two male donkeys. The next year we added a farm pond. After a couple of more dogs showed up and we had adopted four cats, we decided the donkeys could use some friends that actually produced something. We settled on llamas for their regal appearance and their warm fleeces. Beehives soon followed and our garden kept expanding with the addition of a hoop house. Many other animals have since come and gone.

We now have a cut flower business and we sell our flowers, vegetables, and crafts at three local farmer's markets and over the Internet. It's taken almost ten years to get to this point and by any definition, we're still hobby farmers. An off-farm job pays the mortgage and health insurance, and contributes to our savings and entertainment budget. But the farm business sustains itself at a profit, both financially and ecologically. Many times along the way, we were tempted to "compete with the Joneses." But we always reminded ourselves that we didn't need the farm to support us right away and enjoying what we were doing was much more important to us than having the biggest stall at the farmer's market.

Hobby Farming Rules for Success

Low Cost—Always look for used equipment or different ways of doing things around your farm that won't burden you with big bills. If you truly want to run your hobby farm as a small business, you need to make a profit eventually and expensive equipment will keep you from doing that.

No Debt—If you can't pay for equipment in a year, it's not worth having. Downsize your expectations and goals and don't be a slave to the banks.

Good Enough Is Good Enough—Don't let perfection be the enemy of the good. Get the job done and move on.

Sustainability—Always work toward making the different players on your farm—plants, animals, ponds, insects, people—into a sustainable system that's mutually beneficial for all of them.

Organic—You don't need to be certified organic, but practice organic methods in all that you do, from gardening to growing your business.

Quality of Life— The most important element of all is your happiness. Why live and work on a farm if you're not loving almost every minute of it? Weigh all new farm venture ideas against how it will affect your life and relationships. Take the time to enjoy what's been created for you and what you've accomplished yourself.

CHAPTER 2

The Search for a Farm

We understand how you feel. We know why you're reading this book. You want to live the country life and enjoy the peace of mind it has to offer. You can imagine yourself enjoying breakfasts of eggs from your chickens and asparagus from your garden and honey from your own bees on your toast. In the winter, you see yourself snug and cozy by your wood-burning stove, warmed by wood from your own forest, that you gathered and split. You might even be wrapped in a knitted blanket made of wool from your llamas or alpacas. But you've got a long way to go before you get there; first things first, you need the farm itself.

Where will your farm be located? To determine this, you'll have to think about many more things than the sun and the soil. You'll have to imagine all the possibilities that a location might offer you. Is there sufficient land that's even enough—not too steep—to grow crops? How much sun exposure do all parts of the farm receive throughout the day? Are there fields of healthy grass that would support grazing animals or areas that you could clear for that purpose? Is the soil heavy clay or sandy grit? Are there any water features that might help irrigate a garden or help raise the water table near your farmhouse well? Is there enough forested land to manage a wood lot? As we talk about a sustainable hobby farm in this book, you'll see that there are many contributing factor to maintaining a healthy balance on a farm, including water, grass, animals, wood, and your own human spirit. Instead of thinking of your farm as a location, think of it as a living system.

When I asked the owner of the farm we eventually bought if I could come out one weekend to mow his lawn (well before we'd closed our deal), he almost choked on his coffee. I explained that I just wanted to get a feel for the place. It had about three acres of grass around the house that needed mowing and I wanted to see how long it would take. Also, the idea of driving in a circle for a few hours around the farm and viewing the place from many different angles seemed like a good way to check things out.

That one afternoon spent mowing, visiting and drinking a few beers with the owners, provided me more information about the farm than any detailed real estate listing. I noticed it was very quiet (when the lawn mower wasn't on) and there seemed to be no planes flying overhead. I heard a train pass nearby, but it was far enough away to create a pleasant rumbling in the background instead of teeth-chattering noise. I noticed that the grass in the yard and field, while overgrown, was diverse and would easily support many animals. There were few rocks as well, which would make for easy turning of the ground. One outbuilding was used as a shed in which I could envision my own tools. The other was an open barn for the owner's horses to use at will; it had an area cordoned off with wood to store hay. I also stayed long enough to meet the neighbor and realize we could easily get along.

Finding the right farm is both exciting and stressful, but it's not harder than any other house search, just different. You're making a lifestyle change, and because of that, it might feel like a much weightier undertaking. We recommend that you take extra time in your search, even if you have to rent for a while. Because it takes so many years to really work a farm into your own liking, making the correct choice initially is very important. And depending on how quickly you need

to get settled, you're limited by the actual farms that are available at the time you're looking.

Generally, the further away from an urban area you get, the more affordable the property values become. But living further out means a longer trip to the grocery, the hardware store, and especially the markets where you might want to sell the fruits of your labor. Because you'll likely have an off-farm job, you'll want to consider the length of your daily commute as well. If you're an aspiring hobby farmer, some convenience to creature comforts is probably important to you. Today, you can find a location that takes advantage of rural solitude and urban comfort if that's what you desire.

Renting or Leasing

You might not be able to afford to buy right away or you may decide to get your feet wet by renting or leasing land initially. This could be a good alternative, but keep in mind that the terms of the lease and the wishes of your landlord will influence what sort of hobby farming you can do. You may not be able to keep certain animals because the owner doesn't want you to construct fencing. Or much of the land may be leased to other farmers for cattle, hay, or hunting.

On the flip side, the farm may already have equipment for you to use or you may be able to trade farm labor or a portion of the crops you produce for the rent. You'll also not be worried with property taxes or other legal issues of owning. Some longtime farmers who would like a younger farmer to take over their operation may even be willing to offer you seller financing. In this case, you'd be paying the owner as you would a bank and the land would eventually be transferred over to you.

If renting or leasing is your only option, then by all means, go for it. But taking ownership of your land and depositing your blood, sweat, and tears (of joy and pain) in order to enjoy their fruits many years down the road is one of the true joys of hobby farming. And in most instances, it takes years to get your soil in top shape. It would be a shame to work your soil into a nice, black, loam only to be evicted for more development. So moving towards ownership should be your goal.

Location

We began our search by looking first for a small city or urban area that had all the things we'd enjoyed in other places we'd lived: good restaurants, local food and wine stores, farmer's markets, a thriving music scene, a moderate climate, and a progressive attitude. Because our outside source of income comes from a flexible, work-from-home sales job, we had a wide geographic net to cast. Most college towns offer a local food scene that will be ideal for the time when you're ready to get your feet wet at the farmer's market. We decided on the area around Charlottesville, Virginia, for all of these reasons.

Realtors and Other Resources

We signed on with a local Realtor and began the hunt for a house in earnest. We used Realtor.com® to find listings on our own and we scoured the local newspapers (both daily and weekly) for possibilities. The Internet is invaluable in the search and you should complement any efforts of your Realtor with your own searches. Realtors are most useful in navigating the detailed process of finding a place and taking it through the many steps to closing, but they aren't a substitute for your own intuition or tastes.

Keep in mind that many independent-minded farmers won't list their properties with a Realtor and may be selling their properties on their own. So driving the back roads on the lookout for "FOR SALE BY OWNER" signs is another important tool in your search. You may also find useful leads on bulletin boards at the local farm supply shops and the post office.

How to Look

Land in and very near to the bustling center of Charlottesville, Virginia, revealed itself to be rather pricey so we expanded the radius of our search. And with every possibility that opened up, we judged the distance to the city to weigh the costs and time involved in a commute for work or errands. The farther we searched from the city, the more houses with land came onto our radar. Living in the same county where Charlottesville is located was still too expensive for us. But just across the county line, we had more options.

After about a dozen showings and endless hours online searching listings, we found a classified ad in the local paper. The property was just across the county line where the property values were much more reasonable. It was in an area that

was growing and within a twenty-five-minute drive of the city. There was already a farmhouse, two outbuild ings, and fencing for large animals. The house was in need of work, but had a solid structure and was immediately livable. If we'd had kids, we may have also needed to consider the school system. Even if we had not taken the sale through to contract, we now had a good idea of where we needed to look and what we wanted.

Finding the Silver Lining

Once you think you've found a farm possibility, then it's time to begin weighing the trade-offs and looking past the negatives to the possibilities. Every farm or previously owned house has its share of what some might consider problems and others might consider opportunities. It's all a matter of your price range as to how many of those trade-offs you will eventually have to make. We were happy to find a farm within twenty five minutes of the city, but less thrilled that it was accessed by a road running through a neighbor's property and he glowered menacingly through his ZZ Top beard from his trailer's porch each time we drove down the drive. We loved the small stream running through the property, but were disappointed to see that what was called a small pond was really just a mud flat and that there were about fifteen acres of forest that had been selectively logged a few years before, leaving piles of dead debris. The house was structurally sound, but had only one bathroom that would eventually need to be completely replaced, along with a good amount of other cosmetic work. And our immediate neighbor (whose in-laws had originally built our house) was just over a fence, as close as your typical suburban neighbor.

Any of these issues could have been a deal breaker. Clearly they had turned off many others before us, as the property had been on the market for over a year and the price had been reduced. Unless you have deep financial resources and endless amounts of time to look, you'll probably not find your ideal farm. But another true joy of hobby farming is finding a spot, perhaps the ugly duckling of farms, recognizing its potential, and working to reach it. If you don't want to work for it, then why farm in the first place?

We were young and eager to make the place our own, and the price was right. Soon we were fast friends with the surly-looking neighbor in the trailer; we'd made plans to dig an actual pond to replace the mud pit; we had friends over for bonfires to burn the logging waste; we took out a loan to add another bathroom; and we introduced ourselves to a neighbor who turned out to be the kindest, quietest, most helpful person you'd ever want living just a stone's throw over your fence. If we hadn't had the desire to put in the work to make the place our own and fix its problems, we would never have found a suitable farm in our price range.

Tips for your Hobby Farm Search

Choose a livable city or town first and search in an ever-widening radius around it. Use a Realtor from one of the neighboring rural counties, as they will be more tied into farm properties than a Realtor from the city. Search properties on Realtor.com®, Craiglist.com®, Unitedcountry.com®, and in classified ads from the local daily and weekly newspapers. Drive all the back roads around the area looking for ease of access, traffic, desirable terrain, and "For Sale By Owner" signs. Look on the bulletin boards of the local feed and supply stores. Introduce yourself to the owners or managers of the local stores and ask them for leads. The possibility of a new, loyal customer is all the motivation they need to point you in the right direction. Keep an eye out for fixer-uppers. Most of the time, the cost of improvements will not be as much as the money you'll save in the overall price.

Once you've found a potential place to settle, go there often, at different times of the day and night, and just sit and observe. Is it next to a railroad? Is there a large (and smelly) cattle operation next door? Can you hear the highway from your front porch?

Open your mind to the possibilities of your hobby farming venture. If you've found a place that seems almost perfect, but doesn't allow for a specific dream of yours, you can revise the dream. Don't bring a preconceived plan of exactly what you will do on the farm. Let the land tell you what will work best. **Don't settle if it doesn't feel right and don't get too emotional if it does.**

Evaluating a Farm's Water Profile

Think of your farm as a living, breathing organism that needs balance to sustain itself. Water is the lifeblood of the healthy farm. And just like your own blood, it's wise to use and conserve every drop of water that you can. But you can also drown in water, so depending on the climate of your part of the country, you'll either be looking for how to conserve water or how to properly drain it.

When evaluating a farm's water system, you'll want to determine both the quantity and quality of the water. If water is only available by being trucked in and deposited in a cistern, then your water system isn't sustainable. If the fields are flooded each spring, threatening your house, then you've got too much of a good thing. You'll also want to test your drinking water source for quality and your waste water system for reliability.

Evaluating a farm's water profile

- Ask the owner what water issues they've had in the past, but keep in mind they have an interest in playing down the negatives.
- Ask the neighbors the same, but keep in mind that water issues can change from one farm to another.
- Look around the farm for full or drained ponds or creeks.
- Check for erosion, which could signal periods of heavy water flow followed by dry periods.
- Contact the USDA Water Quality Information Center, which tracks drought and water quality for each state and can even pinpoint your particular county: http://wqic.nal.usda.gov.
- Determine whether your property is located in a flood zone by visiting FEMA's website at www.fema.gov or by purchasing a flood certificate from a third party provider.

Zoning, Property Rights, and Easements

As long as your farm is zoned for agricultural use or as rural property, which you can find out from the Realtor or by checking with your county tax assessor, then you should have no problems hobby farming. And since you're most likely looking for property in the country and not a city, then this should not be an issue. You should check with your county office about exactly what is and isn't allowed before you buy.

When you are buying a farm, you are also buying many different rights. Think of them as a basket of eggs. Each egg is a right you are buying with the property. Water rights, mineral rights, royalty rights, surface rights, petroleum rights, even the rights to the air above your farm (for wind power generation for instance) may already be bundled together for you when you buy. In some circumstances, some of these rights may have already been sold to someone else. The title research process, which is required by law when buying any home, is usually arranged by your Realtor or attorney. A title company researches all the previous sales and uses of the property to determine if any of your property rights have ever been sold. If you aren't using a Realtor, then you'll want to hire a title company to do this research for you.

Mineral and petroleum rights can cause the biggest headache. If those rights belong to someone else, then that someone else is allowed by law to come onto your property and mine or drill for petroleum. Imagine waking one day to find heavy machinery blowing rock out of the ground and hauling it away. Worse, imagine if an oil rig is permanently positioned fifty yards from your back porch.

In recent years, there's been a land rush of sorts by natural gas companies to lock up gas leases. Speculators visit farms and try to buy the gas rights or lease them for as little as possible. Many unsuspecting landowners have signed on, only to then realize they are getting mere pennies compared to the massive income generated by the gas wells. They also have their land cut by roads large enough to handle heavy machinery and razor-wire fences constructed around the pumps on their land. Before you buy a property that already has these rights sold, or if you are approached by a speculator, you should immediately talk to a trusted attorney and be very careful.

Similarly, an easement provides another entity with rights to your property that don't include actual ownership. Most commonly, the utility companies have easements under their power lines and they are allowed by law to maintain those easements by cutting trees and, in many cases, spraying herbicides to keep down brush that allows them access to the lines. Or there might be a roadway easement through your property on a shared road as access for your neighbor's property (as we have through our neighbor's property). There might be a public easement owned by the county or state that allows access by the public to a waterway on your property. Or you could set up your own conservation easement to protect your property from development for future generations and perhaps lower your taxes at the same time. Again, the title process will reveal any easements you should take into consideration before buying.

Taxes

Taxes in the country are generally lower than in the city. For a working farm, you'll want to be concerned with property taxes, transfer taxes, income taxes, and sales taxes (sales taxes are discussed further in chapter 12).

In most states a working farm can be eligible for reduced property taxes, the amount of which is decided by the county. This is referred to as an agricultural use exemption or open-space exemption and you can find out how much you might save by contacting the local tax assessor's office. Each locality is different, but the requirement to qualify for a reduced tax bill is that you are engaged in an *agricultural pursuit*, defined as raising a minimum number of animals (each locality has different minimums) or growing crops for sale. *For sale* being the operative phrase. Keeping farm animals as pets does not qualify under the law. Other ways to qualify for an agricultural exemption are to raise timber to the specifications of your county or lease your property to someone else to farm or grow a minimum number of acres for the purpose of selling the resulting crop.

Some states are now even allowing conservation exemptions if you commit yourself to enhancing the environment. Do not think this is an easy out. Conservation exemptions require you to hire a biologist to come up with a conservation plan (i.e., wildlife habitat restoration). Then you must expend the time and resources to execute the plan. Sometimes it is cheaper just to pay your property taxes. The idea is that you are using your land as a farm business or for conservation, and not for development or just your own private enjoyment.

Once you establish your land-use status or buy a property with the status already established, you must stick with it. If you stop farming or you buy a farm and don't continue to farm it, you'll not only lose your agricultural use exemption, but will have to pay more because the government can recalculate previous years' taxes—up to five years—without the agricultural exemption. Talk to your agricultural agent and find out what is required in your area to maintain your land-use status.

There may be ways to buy time when you first move to a farm to keep your land-use status while not farming immediately. We know a farmer who bought a farm that had been overgrazed and overpopulated by cattle for many years. The county allowed him to rebuild his pastures for several years before putting any animals on it, while still keeping his agricultural exemption.

Insurance

We never thought about needing more than homeowner's insurance when we moved to the farm. And we didn't have more until recently. But we began hearing stories about farmers who had cattle escape their fencing and wander into the roadway. We heard of more than one instance of a cow or horse being hit by a car and causing a person bodily harm. The farmers that didn't carry extra insurance to cover this (homeowner's insurance won't) lost their farms in the resulting law suits.

Along with your homeowner's insurance, you will need a farm rider that will cover accidents caused by your farm animals. If your animals are only pets, you may not need extra insurance. But you'll definitely need it if you are selling animals. And if you are growing and selling food at the market, you'll want to insure your farm business from the possibility that someone might get sick or hurt by your product. Talk to your insurance agent for the best approach, but most extension agencies now suggest carrying at least $1 million in liability insurance.

RIGHT: You can contact your local utility and arrange to have your property protected from herbicide sprays that are regularly used under power lines. In most cases they will comply with your request.

CHAPTER 3

Your Farm

Where to Get Information

Not too long ago, there were few resources for a beginner farmer outside of the guidance of another farmer. Still, much of the best advice and instruction you can get comes from a neighbor or farmer you've met at the market. Farmers carry generations of wisdom in their heads, much more than any book can completely cover. And farmers love to share their wisdom. Ask questions when you have their ear and ask farmers you know if you can come out and see how they're doing it.

When it comes to farming, there's no such thing as a stupid question. Every time you visit a farm to pick up hay or buy starter plants, ask the farmer what he or she has going on at their farm. You'll most likely get a tour and you can see how someone else approaches growing crops or caring for animals. You can ask questions and get ideas. Attending farm tours and seminars, which are held periodically through your local agricultural extension agency, is another terrific way to get ideas and to meet farmers that can become a resource for you. You can find a wealth of information in old and new information technology.

You might also begin by interning on a farm or just offering to help out. But don't expect to be paid much. In most cases you might be paid in food from the farm, which certainly has its value. You can look around the larger operations at a farmer's market and feel out the owners. But be prepared to do the grunt work like weeding and spreading mulch. Most farmers don't have time to offer daily outdoor schooling. You can find intern opportunities all across the county on the *Appropriate Technology Transfer for Rural Areas* website—www.attra.ncat.org. This website also offers many valuable resources and articles on all aspects of sustainable farming.

One of the most important resources for answers is an extensive farm library. We have shelves of books on every aspect of gardening, homesteading, and farming. We read up in the winter months, planning our next season. And we go to books when a problem arises and we need answers. There is no single book that will provide you with all the answers; not even this one! You can find an extensive recommended reading section in the appendix with a list of the books we find most useful.

And although high-speed Internet seems like sacrilege to many folks trying to reconnect with the land, the web holds a vast amount of detailed instructions and video for almost anything you'll ever dream of doing on your farm. Satellite Internet now provides high-speed access for even the most remote farm. The initial equipment costs about as much as a new television, but the monthly bills are comparable to regular high-speed access and well worth the expense, in our opinion.

Apply your skeptical farmer filter on all the free advice you'll find online (or offline, for that matter). There are about a dozen ways to do anything and you'll need to weigh each option to find the one that

LEFT: While it's not pretty, this Internet satellite dish is invaluable in obtaining answers to our many farming questions.

best suits your situation. But the Internet is an added expense that may not fit into your farm budget. In that case, use the computers and the books at your local library.

An unlikely source of valuable information is farm and garden catalogs. Once you start ordering supplies and equipment, you'll get more of these catalogs than the ecologically-minded folks behind them should feel comfortable sending. But in those catalogs are ideas for irrigation, greenhouses, natural fertilizers, and organic pest control. And the companion websites for these catalogs often offer much more in the way of instruction and ideas.

Wells, Springs, and Cisterns

Unless you live close to a municipality, you'll not have treated water and sewage. In the country, away from city services, a drilled well is the best water source you can have, outside of a glacial mountain stream. You pay nothing to the utility company for the actual water you use, only the electricity it takes to pump the water out. And drilled wells are deep wells that are less likely to become polluted or to run dry. Drilled wells are expensive, in the range of $5,000 to $10,000 or more, depending on how deep you have to drill to hit water (sometimes 300 to 400 feet). But it will be the best money you ever spent if you are in an area that is prone to drought or you have intensive industry or agriculture around you that might affect your water quality. In most areas, you'd recover the costs of a drilled well in about

LEFT: If your well looks like this, it's most likely a drilled well. If your well is covered with a large concrete cap, it's a dug well. This well should have something covering it to protect it from degrading in the sun.

five to ten years of not paying a water bill, and your water will be free from chlorine or other chemical additives.

A dug or shallow well is the way most wells were created before there was much concern with water pollution. They are shallow, prone to run dry, and the water is so close to the surface that squeezing the water out of the dirt around your gutters might be cleaner.

There are other types of wells and water sources—bored wells, driven wells, or wells that tap a spring. Many times these feed into a large cistern in a basement that holds water for periods when the water source might be running low. All are more prone to pollution and running dry than a drilled well. If you're buying a farm and you don't live in a remote area of the country away from pollution, then spend the extra money and drill a well.

Maintenance of Wells

If you are supplying your own water without the aid of your local utility, then you will need to create water pressure from your well to your house. So you will most likely have a water pressure tank. You'll also have a pump in the bottom of the well. If your well stops pumping, check the pressure tank first and then have the well pump checked. The usual cause of most well problems is with the tank and the pressure switch. It's at most a couple hundred dollars to replace either, so it's advisable that you have a professional come out and do it for you. There are two maintenance tasks with wells that you should perform each year.

Checking Pressure

1. Cut off the power to your pressure tank.
2. Drain the water from your pipes by running several faucets until they stop.
3. Check the pressure in the tank with a tire pressure gauge. There will generally be a pressure outlet at the top of the tank.
4. Follow the directions on the tank for the switch settings you have (the tank cuts on at a certain PSI and off at a higher PSI). If you have no directions, call a service technician. Don't guess.

If you do have directions about the proper pressure that you need, then you can add pressure using an air compressor. Again, follow the instructions on the tank exactly and if you have any questions, call a professional. The pressure should be set at 2 PSI below your pump pressure switch cut-in point.

If you're not comfortable doing this yourself, call a professional, who will usually do it for $50 or less.

The second is to shock your well clean once a year. Even a drilled well can become slightly contaminated. Chloroform bacteria, harmless to most people, can build up and may even be in your pipes and not in the actual well. But the presence of chloroform bacteria can signal the possibility of other more harmful bacteria. A test to detect the presence of chloroform is usually required by law when you buy a farm. It's best to test every few years after that. But if your water tastes a bit off or if it's causing stomach problems in anyone in the household, you'll want to shock your well clean with bleach. Once a year or once every two years is usually enough. Where we live in Virginia, we've done it once every couple of years and never had issues. After you've shocked your well using the procedures below, you can have your water tested again to make sure you've fully cleaned it.

Disinfection Procedures

Use one gallon of regular (no perfumes: i.c. "lemon scent," etc.) household bleach that contains sodium hypochlorite. **WARNING**: There are new household bleaches on the market that do not contain any sodium hypochlorite. Read labels carefully and **look for bleach that contains sodium hypochlorite as an ingredient.**

1. Mix the bleach in five to ten gallons of clear water. Pour half of it directly into the well. If possible, pour the remainder in so that the sides of the casing are washed down. To fully distribute the bleached water in the well, run a garden hose down the well and let it run for five to ten minutes. This will circulate the bleached water from the well through the hose and back to the well again.

2. After the bleach is added, turn on some taps in the house until the odor or taste of bleach can be detected, then turn them off. Do the same for all other taps and faucets. If you have a treatment system for your water, you may need to bypass the treatment unit so it is not damaged by the bleach. If you share the well with someone else, let them know you are disinfecting the well so they do not inadvertently do a load of laundry or take a shower with the bleach water.

3. Allow the bleach water to remain in the well and pipes overnight or for twenty-four hours if possible. Turn on the outside taps (to avoid overloading the septic system) until the smell of bleach can no longer be detected. Then do the same with the taps inside the house.

4. After about one week of use, submit another sample to the laboratory for analysis.

Well Construction Features

A well constructed as follows will help improve water quality:

- A tight cover with a screened vent is needed to keep contaminated surface water, debris, and insects from entering the well.
- The well casing should rise at least six inches above the ground surface to protect it from flooding or ponding of water.
- A watertight casing with solid joints should be sealed on the outside with bentonite clay or cement to a minimum of eighteen feet below the ground. Finding a copy of the driller's well report will help confirm the proper construction of your well. Your county clerk may have drillers' well reports on file; otherwise contact the driller or previous owners.
- The well should be located at least 100 feet from any source of contamination such as a septic drain field, privy, manure storage and spreading, stream, or lake.

Septic System

You may not think of a septic system as environmentally friendly. But if you have a properly installed and serviced septic system, then you are treating your own waste and reincorporating it into your land without the use of harmful chemicals or disposal issues that come with wastewater treatment plants. Environmental problems you've heard relating to septic systems are from older, deteriorating systems that need replacing or have not been properly maintained. A septic system is the most common way of treating wastewater on farms. More people nowadays

use composting toilets as well and recycle bath and dish water (called gray water). If you're starting from scratch, composting is perhaps the most environmentally friendly option and also the most economical. But most pre-owned houses in the country have a septic system.

The main concern is not to adversely affect your groundwater or that of your neighbor. Government regulations are very strict in this area (rightly so) and when you're buying a house, the inspection process should determine if there are any problems. If you're installing a new system (a very expensive process that can run into the tens of thousands of dollars), the same government regulations and required inspections and permits should protect you from any problems.

A septic system takes everything that goes down a drain in your house and collects it into a big underground container. There, it breaks down into liquid and migrates through another pipe to a smaller container. When this container fills with liquid, the waste flows out into very long, thin pipes with perforation that spread out like a fan for dozens of feet away from your house and the septic tank, over what is referred to as the drain field. The liquid is slowly drained through the holes all along the pipes and incorporated into your soil. It's very nutrient-rich at this point and is why you will see very green grass growing over your septic system.

Generators

The first piece of unsolicited advice anyone who hasn't lived in the country will give you is to buy a generator because the power goes out all the time in the country. We've not found this to be true. Sure, there are times during ice storms or high winds when power can be knocked out, but it's never reached the level of the "brown outs" we endured in Oakland during the summer of the California energy crises.

Keep your septic system in proper working order:

- Know where your septic tank is by finding it on the blueprints of your house or the plat of your property.
- Don't build or drive over your septic tank or drain field. You can damage the pipes underground.
- Don't use a food disposal in your kitchen sink. This puts many forms of bacteria into your septic tank that don't break down easily and can result in your septic tank overflowing and clogging up. Compost your food instead.
- It's recommended by septic manufacturers that you use specially formulated bacteria treatment in your tank about once a month. We do not put any food down our drains, so we've never treated ours with the bacteria

and we've never had a problem. But if you are putting food down a disposal, you need to treat your septic tank. This is an added monthly expense that makes composting even more attractive.

· Never put bleach or other harmful chemicals down your drain. Bleach kills the bacteria that are required to break down your waste, thus resulting in clogging and overflow.

· Have a trained professional come out about every three years to pump out your septic tank. They will also use high-pressure water jets to clean out the lines in your drain field. This usually costs several hundred dollars for just the basic service.

· Don't attempt to plant a garden over your septic tank or drain field. The nutrients are too rich to maintain plants properly, and deep-rooted shrubbery or trees can damage the pipes.

In the aftermath of Hurricane Isabel that blew all the way up the East Coast in 2003 and knocked down thousands of trees and our power for four days, we got along just fine without power. We actually looked forward to a few days where neighbors might connect by spending more time outside on their porches. But our hope of peaceful evenings under candlelight and the starry sky was lost with the sound of gas generators running all over the county, like the sound of every neighbor mowing his or her lawn at the same time all day and night long.

Part of the joy of hobby farming is getting closer to the natural world and not relying too much on any one source of food, water, or energy. We lost power last year during a December storm that brought more than twenty inches of snow. We soon realized that our wood-burning stove was adequate to keep us warm. Because it's a true stove, we could also cook full meals on it using the canned tomatoes we'd put up over the summer and the vegetables and pasta we had left in the refrigerator and cupboard. We could have melted snow on the stove as well if we'd run low on water and even enjoyed a hot bath the old fashioned way. We had lots of candles and oil lamps and a house full of books to read. Had we been concerned about the food in the refrigerator and freezer, we could have moved it out into the snow to stay cold.

But even in the summer, is the food in your refrigerator and freezer really worth the cost of a $2,000 generator that you'll truly need only a handful of times? Generators are used so infrequently that it's not uncommon for them to be out of commission at the very time you need them most.

We don't own a generator and don't plan on it. We're confident in the amount of resources we generate for ourselves to be able to survive for as long as it would ever take for the power company to restore service. A couple of days? A couple of weeks? No problem. A few weeks of forced exile from e-mail and television would be a welcome vacation.

That said, if you have medical issues that would make it dangerous for you to be without power for any length of time or you are running a greenhouse full of hydroponic equipment, then by all means, invest in a generator. Another good use of a generator is if you need to run a pump from your pond to bring water up to your garden. But don't run it inside your house or outside in an unventilated area (like in the garage). That's like running your car in your closed garage, which can quickly lead to carbon monoxide poisoning.

Heating with Stoves

Nothing will warm you up faster than an old woodstove, burning the wood you've cut, seasoned, and chopped yourself. The first fire of the season is an annual pleasure equal to the first tomato sandwich of the summer.

Creating your own heat source—if you need one, of course—is one of the true joys of hobby farming. It's also a comfort to know that if something like a massive snowstorm that knocks down trees and power does happen, you'll be perfectly safe and able to heat your home and family. And if you have a true stove, then you can also cook on it.

There are many types of wood-burning stoves, but basically, you pay for the level of efficiency you want. If wood is hard to come by where you live, you'll want the most efficient stove you can buy, a ceramic stove that circulates the hot air. Or a

pellet stove. But a pellet stove is subject to the market for pellets, which can become quite expensive at certain times of the year. And if self-sufficiency is your goal, having to buy pellets isn't a step in the right direction.

You should assess your needs and resources to figure out how much you truly will need a stove to heat. Will it be supplemental heat, like ours? Or is it your primary source of heat? If it's your primary source, then we'd suggest buying the best and most efficient stove your money can buy. Burning wood is not clean and if you're doing it all day long for a winter that lasts five months, you're doing

RIGHT: Our stove is made of steel. It's not the most efficient stove, but it is the most versatile as it can be used for warmth or cooking.

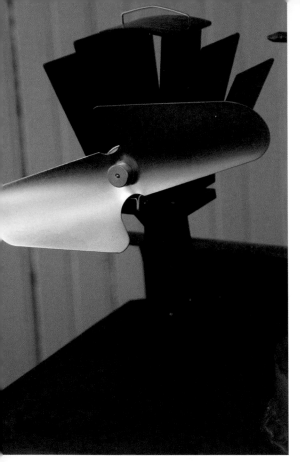

more damage to the environment in emissions than a propane furnace. But diversity of energy sources is a goal of most hobby farmers and a wood-burning stove provides that.

If your farm comes with a stove, we'd suggest that you have a professional come out and inspect it. They can fill you in on any alterations you may need and can suggest ways to keep it clean and operating safely and effectively. If you're buying new, rely on the dealer to provide you with instructions for safe and effective installation and use. This is one area where you shouldn't be reluctant to rely on the expertise of others. Converting old chimneys with stove inserts is also common, easy, and will create a cleaner and more efficient source of heat.

After proper installation, you'll want to practice regular maintenance. Cleaning a stove is a fairly easy and dirty affair. When wood burns, creosote builds up in the chimney. It's the black, oily material that coats the inside of the chimney and the stove. Woods with high sap content, like pine, create more of this buildup. That's why it's always best to make sure wood like pine is dried completely before burning. You can still burn this wood, especially if it's all you've got, but you need to keep a close eye on the buildup in the stovepipe. If creosote is allowed to build up too much, it will heat up, catch fire, and can literally melt through your stovepipe and start your house on fire.

Depending on how much you use your stove, you should clean it at least every four to five months that you use it. And always clean it before the first use in the winter. You can buy a wire brush on a long expandable pole at your local home-improvement store to scrape the buildup off the inside of the pipe. Make sure you also use a wire brush to clean out all the openings in the actual stove.

Farm Safety

Fredericksburg, Texas, where Michael's mother Maryneil and her husband Tom raise longhorn cattle and horses, is a close-knit hobby farming community. Many of the neighbors are retired or close to it. It's a lovely community, full of people who will pitch in for one another when needed, feeding and caring for each other's animals and sharing the burden of large farm projects.

Not long ago, Maryneil and Tom were away on vacation and their generous neighbor was watching their place. The neighbor's washing machine was out of commission, so she decided to bring her laundry over to Maryneil and Tom's place to do her wash. Though they're next door to one another, there's a good quarter mile between their front doors. The neighbor hopped on her four-wheeler, bundled her sheets on the handlebars, and set off.

As she was zooming along, she didn't notice that one of the sheets got loose and tangled in a front wheel. The four-wheeler instantly flipped tail over-head and catapulted our friend into the field. Her husband saw what happened and ran to find his wife in a heap with a broken hip and shoulder. She's quickly recovering now, as most hard-headed farmers do, but the episode could easily have become fatal.

It's too obvious to say that farms are dangerous places. There are loads of statistics to back this up. What you need more than anything else to avoid injury to yourself or your animals is common sense, patience, and intuition. But even the most seasoned farmer can make mistakes.

When we began working our farm, we had a sense of urgency to whip it into shape. We worked furiously digging vegetable beds, stringing fencing, and mowing grass and brush as tall as we were. After a back injury, a bout of poison ivy, and countless cuts, scrapes, and bruises, we finally realized that there is a safe pace to do farm work and an unsafe pace.

This may sound trite, but having talked with many new farmers, the proper pace of farm work is almost a universal lesson to be learned. When you have a big project in front of you, never look to the end and race towards it. This is a recipe for injury. Envision all the separate steps that it will take to get the job done so as not to overwhelm yourself and work only on the current step at a slow and deliberate pace. Farm work is meditative. It's when you're most focused and steady in your work that injuries are avoided.

A good way to put yourself into this "farm rhythm" state of mind is to chop and stack a large pile of firewood. We'll discuss the proper techniques later in that section. But the idea is to have a large pile of wood and to slowly work into a rhythm and keep it up until you reach the end. Don't swing wildly at the wood. Slow and steady wins the race. Don't get angry at the knots. Just realign the wood and try again. Set aside the pieces that refuse to be busted until another day. Soon you'll be surrounded by a large pile of cut wood, you'll have gained a solid workout for your mind and body, and you'll have experienced the farm work rhythm that you can

The Learn-from-Others'-Misfortunes Section

- A farmer was knocking down low-hanging dead limbs on a tree using the bucket front-end loader on his tractor. A limb a bit higher came off as well, landed on his head, and almost knocked him out while still on the running tractor.
- A woman we know had stalled her farm truck on the top of a cattle guard. She got out to push it off, lost her footing, and fell under the truck. It ran over her leg, but fortunately didn't break it.
- A farmer left his prize horses in a big field with no shelter. A thunderstorm blew in and one of the horses was killed by lightening.
- A farmer who was working next to his electric fence leaned over and brushed his cheek against it. It knocked him down flat and he said it felt like someone had cold-cocked him.

practice on more complicated projects down the road. Now that you're sufficiently tired and calm, don't forget to stack up all that wood.

The other key to avoiding injury is to know when you've bitten off more than you can chew. Don't cut down a tree you're not comfortable handling, for instance. There's no shame in asking other farmers for help. Know your limits.

Guns and Hunting

Without getting into a debate about our constitutional right to bear arms, let's just stick with the question of whether you will need a gun or not on your hobby farm. We've never owned one and don't plan on it. But there are several reasons that farmers do keep guns. Protection, hunting, and compassionate euthanasia of animals are the three most common.

We have four dogs that would give any burglar a run for their (or our) money. And we keep bear-repellent pepper spray that could disable an intruder of either the human or animal kind. We keep our garden and vulnerable animals close by and surrounded by tall woven-wire fencing with our dogs as constant guardians. Our donkeys and llamas keep any predator from even approaching the fence. Guns are just one more danger on a farm that we choose to forego.

Tips for Farm Safety

✓ Wear protective eyewear and earplugs whenever operating machinery, even a lawn mower.

✓ Never operate machinery or a chain saw after having even one drink. Alcohol quickly impairs your judgment.

✓ Wear work gloves and closed toe boots or shoes when doing any farm work.

✓ Avoid wearing loose clothing when operating machinery of any kind.

✓ Keep fire extinguishers in your kitchen and in your barn.

✓ Never operate heavy machinery or a chain saw without someone else nearby or at least at home that can call for help if you need it.

✓ Keep a first aid kit for humans and also one for your animals.

✓ Post all emergency numbers and contacts clearly on your refrigerator in case someone else needs to call for help for you or if something happens to your animals while you are away. And make sure to share your emergency information with your neighbors.

✓ Create proper fencing and regularly inspect it in order to keep your animals safe.

✓ Keep a calm and assertive demeanor when working around animals.

✓ Study the operator's manual of any machine before you use it.

✓ Never walk behind a tractor when the power take-off (the twisting knob on the back of a tractor that powers farm implements) is engaged. Clothing can easily become entangled or debris can be sent flying by cutting equipment.

✓ Always walk your field or lawn before mowing to look for hazards.

✓ Wear a protective facemask when spraying chemicals (even organic ones) or doing yard work to avoid allergy problems.

✓ Get to know your local plants, insects, snakes, and animals so you know which ones to avoid.

✓ Never put up wet hay in a barn. It can combust and is one of the leading causes of barn fires.

✓ Wear sunscreen.

✓ Keep children on a short and painful leash. Well, at least keep a close eye on them and teach them what they should avoid.

Many farmers keep guns for the purpose of preventing misery in their animals. Perhaps it's a Sunday and you've found that one of your horses has a compound fracture of his leg and he is suffering. You're unable to get in touch with the vet, so what do you do? The humane thing to do would be to put a .22-caliber bullet right in the horse's forehead. This is a harsh reality of life on a farm. Sometimes animals end up in unbearably sad situations and it's your responsibility to make sure they don't suffer needlessly. We know enough people that actually do have guns that we'd be able to borrow one at the drop of a hat, so we don't feel we need one for ourselves. And any other day of the week besides Sunday (and probably even on a Sunday in most cases), we'd be able to get a vet out right away to administer sedatives and put the animal down less violently. And keep in mind that any animal in a life-threatening injury situation will quickly go into shock and its body's natural defense mechanisms will dull its pain. But if we lived too far from immediate help, we'd seriously consider having a gun around for just this purpose.

Farm Equipment and Tools

Sometimes it seems that farm equipment and their manufacturers have some secret means of communication. Right about the time the brand new tractors come rolling onto the dealers' lots or the spring sales special flyers reach your mailbox, some piece of expensive equipment inevitably breaks down or becomes tiresome to operate. Don't you really deserve one of the newer, easier-to-use versions? Inevitably if one thing breaks or needs replacement, at least one or two others are soon to follow. And what a joy it is to buy a shiny new chain saw that runs like a top and cuts like butter.

Farm Truck

Our first farm truck was a non-road-worthy, 1983 Datsun with a rodent problem. Audrey refused to even sit in it because of the odor, a unique musk. For several years, we kept it running with pure guesswork and religious conviction. Its now in its final resting place in a remote area of our farm, having become stuck in the mud with a broken clutch line, like a soon-to-be-extinct dinosaur frozen in its last act for future archaeologists to ponder. May it rest in peace.

Our current truck, nicknamed Jack, is a green Ford Ranger with about 150,000 miles on it. The six-foot bed has been made filthy by all kinds of farm detritus. The four-wheel drive has ferried us down our long drive with almost a foot of snow on the ground. Jack has hauled everything from loads of mulch to a pallet of landscaping rocks to a broken lawn tractor. Outside of the chain saw, he's the best piece of firewood equipment we have.

LEFT: Dolly and Tippy waiting for the ride to begin. An all-terrain electric vehicle is invaluable if you've got a lot of land to cover

I recently spent $4,000 to have our old Massey-Ferguson tractor completely overhauled. I replaced the cylinders on the front-end loader, fixed the power steering and brakes, and updated the electrical system. Soon after I got it back from the repair shop, I went to cut the grass in the field with the pull-behind shredder. But the shredder had blown a bottom gasket and wouldn't hold fluid. I decided to try to take it into the shop myself, so I got out my big chain with hooks and attempted to lift the shredder into the back of our truck. It was way too heavy for our little truck and the tailgate immediately bottomed out and severed the brake line. I could have stopped there and just decided not to use anything with a motor for the rest for the season. But soon after I drove the truck into the shop in low gear so I could stop it without working brakes, the chain saw just wouldn't stay running while I was attempting to cut some trees. All the manuals tell you not to attempt to adjust a chain saw carburetor yourself unless you know what you're doing. But how hard could it be? Soon I'd set the carburetor too rich and blew the engine out completely. Oh well, I had my eye on a new chain saw anyway. And for icing on the cake, I was lubing up the lawn tractor (a different tractor from the one I'd only recently spent $4,000 fixing) to get it ready for mowing season when I made the rookie mistake of forgetting to refill the oil after draining it and changing the oil filter. I fired her up, and about thirty seconds later, there was a sound like a gunshot. I had inadvertently put our lawn tractor down for the count. Soon, with about $6,000 total lifted from my wallet (or 24,000 ears of corn sold direct-to-customer), I was ready for spring with a whole barn full of brand-spankin' new equipment. This is why the only profitable farmers are those that learn to fix their own mistakes or avoid them in the first place.

—Michael

While you can easily get along without a farm truck by borrowing from a friend or paying for delivery of whatever you need hauled, there's just no substitute for this essential piece of farm equipment. You don't have to have the biggest truck around. A small truck, like a Ford Ranger or Toyota, can do more work than you can imagine. Of course, if you plan to pull a trailer full of horses, you'll want to consider something bigger. But if your main concerns are going to and from the home improvement store, picking up loads of mulch and compost, bringing in firewood from your forest, and hauling trash, then you don't need a two-ton beast. Four-wheel drive is an almost essential feature, but that all depends on your climate and your land.

Whatever you do, don't buy a dually. A dually is a big, diesel truck with four wheels in the back that costs three figures to fill up. These monsters are for agribusiness farmers and they cost as much as a small house. The quickest way to draw attention to yourself as a farm poser is to drive a brand spankin' new dually up to the feed store and bring home one bag of sweet feed. Better to have your Ford Ranger dragging its muffler on the ground and shooting sparks down the road under a load of soil, than to be that new guy from the city spending his entire farm budget filling up his dually with diesel.

Tractors

Owning a tractor is better than therapy. There's no greater meditation than driving slowly in a pattern around a field, cutting the grass down, or dragging the earth to prepare for a planting. The resulting grid of disturbed and neatly cut ground appeals to something deep inside us, like a geometric ordering of our souls with the universe; a crop circle to the gods. The simple design and reliability of a tractor will swiftly renew anyone's faith in mankind's ingenuity.

ABOVE: Our Massey-Ferguson 65, built in 1963 and still going strong.

Tractors are not toys. The most serious and common injuries reported on farms every year are from tractors. They can roll over or loose clothing can become entangled in the spinning parts. The old tractors can be hard to drive. Our tractor didn't have power steering when we bought it. So it was really tough to turn it, if it wasn't moving. I was cutting grass with it one day around the barn. As I turned toward the barn, I misjudged the space between it and my front-end loader. The rip of the wood through the corner brace of the barn stuck in my head for a week. I was lucky the whole barn didn't collapse as I'd taken out a load-bearing corner post. A trip to the lumber store and much work reinforcing the barn has kept me from trying to cut the grass perfectly around the barn ever again.

When considering a tractor, the idea is to match it to the needs of your land. If all you're doing is cutting grass, then a lawn tractor (otherwise known as a riding lawn mower) might fit your needs. The lawn tractor can be fitted with attachments, such as small trailers and snow blades, ideal for a very small farm of a couple of acres. Resist the urge to buy the biggest, baddest lawn tractor on the lot. They all have the same engines with tiny variations in horsepower, and now they all have cup holders. Mowing your lawn with a 48-inch cutting area isn't any quicker than if you do it with a 42-inch.

If you're planning to do more than mow, you'll need a bigger

LEFT: A three-point hitch and/or power take-off (PTO) are needed to use most farm implements. In this photo, the PTO is the small knob at the middle and bottom of the machine, right above the long draw bar at the very bottom. The two arms on each side of the draw bar are two of the three links in the three point hitch.

tractor that comes with a power take-off (PTO) and three-point hitch that will power all your various implements. Outfitted with the proper implements, a tractor with a PTO can move earth, plant and cultivate many acres of field, harvest crops, cut and bale hay, dig fence posts, level roads, and much more.

New tractors, while expensive, typically have four-wheel drive, which is a huge benefit. If you have four-wheel drive you need much less horsepower to do almost anything. And it's harder to get your tractor stuck in the mud. Even a used tractor built after the 1990s will run you close to $10,000 or more. But in terms of overall ease of driving, safety, and functionality, the newer tractors with four-wheel drive can't be beat and are worth the money, if you've got it.

If your budget doesn't allow for one of these newer tractors or nostalgia pulls you in a different direction, you'll want one of the old workhorses—a John Deere, Massey-Ferguson, or Ford. Tractors retain their value like nothing else with wheels. And there's a good reason for that as they last longer than farmers or farms. We bought our 1963 Massey-Ferguson 65 for almost the exact same price that it was originally purchased new, if you don't count inflation. And its engine still runs with as much compression almost fifty years later.

If you're smart, you'll buy a make of tractor that can be serviced by a local dealer or repair shop. While many mechanics will come to your farm to make tractor repairs, there will inevitably be a time when you need to take your tractor into the shop. You'll discover that tractor dealers and mechanics, like farmers, are few and far between. Our closest Massey-Ferguson dealer is fifty-five miles away—a definite oversight on our part when we bought it.

When looking for used tractors, ask the dealers and sellers a lot of questions and don't hesitate to answer their questions. The sales people will ask you what you're planning to use it for, whether it's cutting several acres of grass, digging holes with a backhoe, or plowing and planting grain. They'll help you figure out the type and size of tractor you need. You can also check your local used equipment classified rags that are usually for sale next to the register at your local convenience store. We have one called *The Buck Saver*, and that's where we've found most of the farm equipment we've bought. And more and more farm equipment is now showing up on Craigslist.com and eBay.com.

If there's one attachment on our tractor we use more than any other, it's the front-end loader. This bucket on the front of tractors does more than move dirt,

Tips for Finding a Tractor

- Assess your needs and ask your local tractor dealers what they've got to match them.
- Look for used tractors first, unless you've got plenty of money for a new one.
- Always try to pay cash or pay off your new equipment within a year unless you don't care to make a profit on your farm in the near future.
- Find your closest tractor repair shop and buy the brand that they service. You'll save lots of time and money this way if you ever need it repaired.
- Buy a tractor with a front-end loader. Adding one later can cost almost as much as the tractor itself.
- Buy a tractor with four-wheel drive if you can afford it.
- Look in your local farm classifieds for tractors that are being sold with many of the implements included.

Common implements that a hobby farmer might use:
Front-end loader – Scoops dirt and lifts other equipment.
Backhoe – Digs holes and trenches. Extracts tree stumps.
Shredder or Bush Hog (Bush Hog is a brand name) – Cuts grass and brush.
Plow – Breaks new ground for planting.
Disk – Evens out the ground made rough by plowing.
Hay fork – Lifts big round bails of hay.
Blade – Drags gravel and evens roadways.
Auger – Drills fence posts.
Tiller – mixes soil and destroys weeds.

Extract Your Tractor from the Mud Using a Front-end Loader

1. If you find yourself stuck in the mud and you have a front-end loader, rotate the bucket down all the way.

2. Lift the front end off the ground by lowering the front-end loader.

3. When the wheels are off the ground, put the tractor in reverse and rotate the bucket back up. This will push the tractor back several inches. You may need to do this several times to work your way back out of the mud.

although that in and of itself is plenty. We mostly use it to collect manure and build and turn our compost piles. You can also use the front-end loader to move very heavy pieces of equipment. Just hook a chain to the bucket and you can lift anything under the payload and move it around or into the back of your truck.

Tools

There is indeed a tool for everything you might do around the farm, from a chain saw to cut a tree to micro-tip pruning shears for delicate cut flowers. But there are also many more tools that can be used as substitutes. And finding something that works well enough instead of buying a brand new tool that you might not use often is a victory of ingenuity that you should immediately celebrate with a beer or a ten-minute break.

Do you need to build a lean-to out in a field for your animals and you don't have electricity nearby? You don't need to buy an expensive generator. Get your chain saw out and do some cowboy carpentry. Just make sure to buy your wood a little longer than you need or adjust your measurements, because wood is going to fly. Never buy a specialty tool before you actually need it (no matter how cool it is), as you're not going to use it. Trust us.

Many specialty tools, like tractor implements or tools used for fencing, will be discussed in later sections dedicated to those subjects. But there are a few universal tools that you should start out owning as a hobby farmer.

Tools for beginning hobby farmers

Tape measure	Work gloves
Drill—electric and battery powered	Skill saw
Heavy flashlight	Headlamp
Hand lamp	Duct tape
Air compressor	Vise wrench
Vise	Bench grinder or angle grinder
Tire pressure gauge	Battery charger
Bottle jack	Come-along
Tie-downs	Tarps
Bungee cords	Nylon rope
Shovels—spade, snow, and manure	Sledgehammer
Wheelbarrow	Staple gun
Saws—various	Level
Triangle	Crowbar—small and long
Wire brush	Pallets
Weed whacker	

Attaching a power implement (in this case, a shredder) to a tractor. Warning: Never attach an implement with the PTO running. Check to make sure the brake is set before moving to the back of the tractor.

1. After backing the tractor up to the implement, use a long crowbar to line the implement up perfectly.
2. Attach the left side draft arm by sliding the hitch pin through the implement opening and the ball end of the tractor arm.
3. Secure it with a locking security clip. Attach the right side the same way.
4. Connect the top link from your tractor to the corresponding top link on the implement. Secure with locking security clip. You may need to restart the tractor and raise or lower the arms to get this lined up just right.
5. Attach the drive shaft of the implement to the PTO on your tractor.
6. When attaching the drive shaft to the PTO, make sure it locks into place.
7. Tighten the top link by inserting a screwdriver into the hole in the middle and twisting it until it's securely tightened. Tighten until the implement is level from front to back in the desired operating position. Again, you may need to raise or lower the arms to get this just right.
8. Twist the leveling crank on the right side of the tractor until the implement is completely level, from side to side.

LEFT: The come-along is one of the more versatile farm tools. You can use it to pull a piece of equipment from the mud, stretch wire fencing, and straighten a building during repair.

Fencing

The call came early one morning from a man up the road. "Do you have two white horses?"

We have two donkeys and both have some white on them, so I guess that's close enough. Turns out, they were on walkabout, a good quarter mile up and on the other side of the highway. The neighbor had called animal control for help, only to be told that animal control will not deal with people's loose livestock. Never mind that they could have been hit up on the highway, causing an accident and/or getting killed. Due to the kindness of this neighbor who asked around to find out who the donkeys belonged to and drove over to pick me up, we were able to track them down, halter them up, and lead them back home.

I'd been looking at the large gap between two wires that had formed in the fence at the corner of our field for months. One middle wire had come loose. I couldn't imagine donkeys squeezing through that gap. But donkeys are a wily bunch and they had done just that. I brought them home, tied them up, and immediately went out to fix that problem. Then I went straight over to the neighbor's house with a twelve-pack of his favorite beer. It's good to have concerned neighbors and better to have fool-proof fencing.

—Michael

BELOW: 1. To build high-tension wire fencing, first drive the t-posts into the ground with a t-post driver. 2. Then run your fencing wire and tighten it securely at the ends with the special cams (make sure you cross-brace the ends and corners with the special brace kits that you can buy with the other fencing supplies. They come with directions). 3. Secure the wire with a fencing tool along the fence line to the t-posts at every intersection with the fencing clips.

ABOVE: Woven-wire fencing can be attached to any posts. In this case, it's used with welded pipe in order to lessen the chance that horses could be injured or hung up on it.

ABOVE: Barbed-wire used in conjunction with woven wire to keep longhorn cattle contained.

BELOW: A hot wire is run out to electrify the fence around a mobile chicken coop.

BELOW: This board and rail fence has a stile constructed in it to allow easy access by people while keeping horses from passing through it.

LEFT: Fencing like this was common in the Civil War era when metal was in short supply.

Regardless of what type of fence you have, if its job is to keep in animals, you should walk the fence line once a week to look for any problems. Wires break from animals leaning on them. Trees in the forest fall on the fence line without making a sound. In fact, there's a gate post rotting away right now in the corner of the field that needs replacing. We should go do that now.

The type of fencing you choose may depend on what's there already. Our high-tension wire fencing was put up by the previous owner who kept cows and horses. Unless we wanted to change or add to ten acres of fencing, then we also needed to stick with larger animals. This is one reason we chose to keep donkeys and llamas and not goats or sheep.

Common Types of Fencing
- High-tension wire with t-posts—used for large animals like horses, cows, donkeys and llamas.
- Woven wire—used to keep smaller animals in or predators out.
- Barbed wire—recommended for use only with cattle. It's very dangerous for any other animal.
- Electric—woven and high-tension fencing; solar, battery, or AC/DC power. Recommended if you want to move your fence line often or if you need that extra deterrent for larger animals like cattle or predators.
- Board and rail fencing—best for horses. It's expensive to install, but offers the least chance of damage and escape.

Gates and Cattle Guards

You can spend half of a work day constructing a gate that fits with your aesthetic sensibilities. Or you can buy a metal gate for under $100 that will last decades and you won't have to worry about it ever sagging (unless you secure it improperly). Lightweight metal gates are easy to transport and use. If it's a large gate, make sure that the post to which you attach it is large enough and concreted into the ground with backup support to a post next to it.

LEFT: This old feed house provides shade for animals and the gaps in the siding allow proper air flow to store hay and feed.

RIGHT: A cattle guard keeps cattle from passing through a gate as they don't like a step on or between the pipes. But don't use one if there's a chance that a horse could get stuck in it.

Avoid This Mistake

A horse escaped a poorly designed enclosure and its legs became caught in a cattle guard out on the county road. After a whole night of struggle, which left the horse crippled and in severe pain, the horse had to be put down. Cattle guards and horses don't mix. Remove cattle guards from your property or make sure your horses are completely isolated from them.

There's no such thing as going overboard when securing the gate post. Years of holding up a heavy gate without moving requires a good amount of reinforcement. Avoid attaching gates to trees. The tree will grow and the gate supports will move with the growth and soon your gate won't open or close properly.

Cattle guards should be used sparingly, and very carefully, if you have horses. If you have several hundred acres and multiple entrances and exits, it might be worth having cattle guards. But these are for convenience purposes only as they save you from having to open and shut a gate.

Outbuildings and Barns

Standing inside an old barn with sun slanting through the cracks, illuminating the manure dust in the air like an old time projector, one can't help but imagine the human and animal lives that have begun and ended inside over the years. Barns see the birth of foals and calves. They hold nests of barn owls that hunt the mice and rats that inevitably inhabit the corners. Restless horses have paced the ground while lightning cracks outside on hot summer nights. Children have conquered countless fantasy empires in battles staged from the rafters. Old

animals have spent months sheltering inside during their later frail years and have taken their last breaths while lying on their sides, soothed by the fading voices of their people reassuring them on to the next life. And yes, barns have sex appeal. They don't call it a roll in the hay for nothing.

Some see an old barn, leaning to one side with rotting slats of wood, and immediately imagine tearing it down and starting over. Sometimes taking a barn down and reclaiming the usable old wood is the best option. Some barns are in such a state of decomposition that it's too dangerous to attempt a repair. But they just don't make barns like they used to (except for the Amish) and building a new one is a very expensive proposition, easily into the tens of thousands of dollars. As the old barns continue to rot and fall, consider it a duty of historic preservation to refurbish one. In fact, you may very well qualify for historic preservation status and you might even be able to secure money from the state to rescue it. Contact your state historical preservation office to explore that possibility.

Our barn is modest, both in its size and its architectural detail. But it's more than seventy years old, older than our parents. So when, after all these decades, we were hit with the heaviest snowfall in a hundred years and the roof buckled under the weight of two feet of wet snow that turned to ice, we never considered taking it down. A buckled roof is one of the most common ailments of an old barn or outbuilding. Dry rot, which leads to the classic "barn lean," is another. These are not unfixable problems and you can add decades to the life of a barn with simple reinforcement. Below, I've chronicled the process of restoring the pitch and shoring up a barn roof.

Farmer Profile

Tom Martin, Poindexter Farm, Louisa, Virginia

Twenty-five years in the building business had run their course for Tom Martin of Poindexter Farm in Louisa, Virginia. Having his own construction company had taken its toll on him physically and he'd gotten out of it all he could economically. So an early retirement at age fifty-five left him with time on his hands and the freedom to do something new and different.

But finding a job or working for someone else didn't appeal to him. As he describes it, "I'm used to getting my way and I don't take orders well, so I'm not the most employable person."

He and his wife Kim, who has a steady job at the local university, invested in a twenty-four-acre farm and Tom began exploring what he might do with the land. He had a friend who had been selling cut flowers at the local farmer's market who saw that there was a market for grass-fed beef. And after attending a conference on berry production offered by the agricultural extension office, Tom decided on berries, beef, eggs, and asparagus as the basis of his hobby farming operation.

"Diversification is the most important thing for any farming operation, no matter what size you are," says Tom. And he's not alone in this observation. Diversification is now one of the cornerstones of sustainable agriculture.

Poindexter Farm can be described as a seasonal grass-based farmer's market business. Tom starts early in the season at the farmer's market with asparagus, blueberries, and raspberries, including homemade raspberry jam. He buys steers and typically has half a dozen at a time, which he rotates onto fresh pasture regularly. The cows are never on the same pasture twice in one year and are sent off to be butchered after about two years. He's built a mobile chicken coop for about eighty hens (producing almost eighty eggs a day) that he moves with his tractor three times a week around his pastures, which have electric fencing to keep predators out and the chickens in. It's easy to see where the chicken tractor has been, as the grass is lush and green, even during the recent drought. Cows will soon be on the pasture that's been improved by the chicken's manure. To keep his costs down, he sells the chickens in the fall in order to avoid feeding them all winter long. He'll order a new flock in the spring.

By all accounts, Tom says that his farming venture has been the best decision he's ever made. He truly enjoys the social interaction he has with customers and other vendors at the two farmer's markets he frequents every week from April to October.

His advice for others who are selling at the market (besides diversification) is to keep your expectations in check and to be realistic with your goals. It takes time to build a customer base and if you're doing all the work yourself, there's only so much you can do, especially when you're starting out.

Tom thinks of the market season the same way you would a baseball season. It's a long season. There will be games that are rained out. There will be times when there are no fans (or customers). Some days you win and some days you lose. But it's the entirety of the season that matters and you'll have time in the winter to make adjustments, trade your unproductive players, and prepare for the next season.

In just his second year at the market, Tom is turning a small profit. Kim brings in a steady income and health insurance benefits that relieve the pressure on the farm business to grow so big that it becomes overwhelming. The work is rewarding and good exercise. His expectations are realistic, and most importantly, he's having fun.

Firewood

Being able to use your land's materials to build structures and stay warm is one of the most satisfying parts of being a hobby farmer. Cutting your own wood from your own land is the cheapest way of creating heat and obtaining building materials, and also has the added benefit of improving your mind and body, if done safely. You'll find that the process of selecting and cutting trees from your own forest for firewood or building material forms an instant connection with our earliest human ancestors.

Even if you mechanize every conceivable method for obtaining wood yourself, by using a chain saw to fell the trees, a truck to transport them out of the woods, and a log-splitter to create your firewood, you'll still gain a valuable workout and more appreciation of the heat and materials you gain from it. After all, you'll still need to be thoughtful and careful in choosing which trees to cut and managing your woods sustainably. You'll need to focus your mind and pay attention to avoid injury. Your muscles will be taxed while wielding a heavy chain saw. You'll push your stamina to the limits loading truckloads of heavy, wet wood. Later you'll haul pieces of that wood onto the log-splitter and then you'll stack it. Of course, the more work along this process that you do manually, the more energy you will need to exert both physically

Tips for Fixing Old Buildings

- When assessing the condition of an outbuilding, the key thing to look for is dry rot. Take a screwdriver or ice pick and stab at the load-bearing posts and the rafters. You'll quickly know whether there are boards that need reinforcement. Most of the time, you can just install another post or board alongside the rotted one and you don't need to remove the rotted board at all.
- Follow the original construction and use metal corner braces to add more support.
- Look for water damage in the rafters and holes in the roof. If the roof doesn't leak, then a barn can stand for a hundred years or more in most cases. Plug the holes in your roof immediately, but be sure that it's stable enough to work on.
- If your building is leaning to one side, you can shore it up at its current angle of lean, depending on the severity. Or you can right it using your tractor or truck and a heavy-duty chain or rope. Be careful to use a long enough chain or heavy rope so that you don't pull the building on top of you if it were to fall over. Tie the chain to the top boards on which the rafters sit (you may have to remove some of the side boards to get to them). It's also best to tie to two different points to provide a more even distribution of force. Then slowly pull the building upright. Once it's straight, reinforce the corners and angle brace the walls to shore it up. But be very careful that the structure is sound and not in danger of coming down on top of you. It's best to remove the outside boards and work on the load-bearing posts from the outside of the building where you can make a quick escape.

1. The roof of the donkey shed bucked under the weight of heavy snow.,

2. We propped the roof up immediately with posts, 2'x6' boards, and concrete blocks to stabilize it.

3. We then used come-alongs and heavy gauge wire to pull in the sides of the structure.

4. We used bottle jacks to raise the roof while alternately tightening the come-alongs until the roof was straight. We then sunk posts supporting the roof peak in concrete to shore it up for good.

and mentally. And the more satisfaction you will gain. Because hobby farmers need to manage their time wisely, you'll need to decide on your own which parts of the wooding process you can do manually and which you'll want to mechanize.

Felling trees and chopping wood are some of, if not the most, dangerous things you will do on your farm. There's no logging police to make sure you wear the proper safety gear or practice the proper techniques, but remember that oftentimes there is no one around when something goes wrong. You should have someone with experience take you out for your first excursion felling trees. Don't be embarrassed to ask a neighbor. That's what we did. In most cases, they will be more than willing to offer you some guidance and help. Who wouldn't want to help a neighbor stay alive? If your neighbor is really busy, offer to share the wood you're going to be cutting.

Managing a Woodlot

There are many philosophies about managing woodlots. And there are different techniques, depending on your goal, whether it be for firewood, wildlife habitat, or building materials. We practice a selective cutting management system. Selective cutting is a method that preserves all different species of trees while only cutting those that are mature or if cutting them would benefit other growth nearby. In this way, you maintain the full diversity of your forest while cutting the trees that will be first to die. This will open up your forest for the smaller trees to get sunlight and grow. You will also cut trees that are growing at weird angles, are damaged, or are crowding other trees.

When talking about managing a woodlot, we're primarily referring to managing it for wood burning. But if you have valuable older trees on your lot, like black walnut or white oak, then you will want to save the straightest ones for possible sale to your local mill, mill them yourself, or just leave them to enjoy in your old age. Also, leave most standing dead trees that have evidence of animal activity (cavities and bird nests). These serve as valuable wildlife habitat. Similarly, dead and rotting trees on the ground create small ecosystems on the forest floor that support fungi and valuable insects and provide natural fertilizer and mulch for the forest.

Use your intuition when deciding which trees to cut and which to leave. The space between each large tree should be around eight to twelve feet. Don't clear out all the young saplings because you'll want to have some new growth. Walk your woods and imagine how the sunlight will penetrate when you take a tree out. It's a good idea to mark each tree you plan to cut with surveyor's ribbon and then come back several days later and reconsider. There's no magic formula, but just remember that it takes decades for trees to reach maturity. Respect them and only take the trees you need and those that will allow others to thrive. If you're not confident in choosing the right trees to cut, then you might enlist the help of an agricultural extension agent who is familiar with forest management.

The trees in forests or on farms vary widely across the country. Our family in the Texas Hill Country has sporadic stands of low-growing cedar and oak. The oak trees have a terrible disease called oak wilt that is killing millions of them. No one knows the cause of this disease and scientists are working diligently to try to breed trees that are resistant. In the meantime, there are large stands of dead oak that pose a serious fire danger. In this case, you'd want to cut out all that dead and diseased wood first. In the Rocky Mountains, the pine bark beetle is creating the same situation. But cutting dead trees is a dangerous proposition and they can fall in unexpected directions and ways, so if you're faced with this problem, you should have a logger or forester come out to evaluate your situation.

Hardwoods, like oak, maple, birch, and ash, are the most efficient woods to burn. But these are also the most valuable and take the longest to grow. That's why in selective cutting you fell a mix of species, including soft woods such as pine and cedar. For this same reason, when you're burning in a wood stove, it's best to mix it up. Start with well-aged, "hot" wood like pine to really get the fire cooking, then add on the heavier oak and maple to cook for hours.

Tools

Chain Saw

How big is your chain saw? (That's a very personal question, but one you shouldn't be afraid to answer.) If you're only cutting a couple of cords of wood each year, along with a few trees that may have fallen over your driveway, a small twelve-inch chain saw is adequate. A small saw weighs less and is easier to handle. (And it won't wear you—or your arms—out). Some of the more recent smaller saws have a dial on the side with which you can tighten the chain by hand. Avoid this gimmick. We bought one of these and have found it to be difficult. It's just a matter of time before the cheap plastic handle will break. Try to find a saw that tightens the old-fashioned way, with a hex nut and flat-head screw top.

Once you get the feel for a smaller saw over a season or two, your muscles will be trained to step up to the bigger models. The medium to bigger saws, sixteen inches or more, are beautiful machines. You'll want one eventually. But save it for when

you really know what you're doing. It'll be that much sweeter when you do finally purchase it.

Maul and Ax

Many a hobby farmer has caused him- or herself a whole lot more time and effort than need be by trying to split a load of wood with an ax. When you're splitting wood, you should primarily use a maul. A maul looks like a sledgehammer on one end that tapers down to a point at the other. It's heavy and rips the wood apart instead of cutting it. Our neighbors ask, "You gonna' bust all that wood?" They don't ask if we are going to cut it because they are referring to busting it with a maul. An ax is used for chopping trees down, if you want to do it the old-fashioned way. Or if you need to split very small pieces of wood, then you'll want to bust it first with a maul, then spit it into smaller pieces with an ax.

Timber Jack

This handy device is used to grab a fallen tree and lever it up off the ground. Once raised off the ground on the timber jack, the tree is easily bucked up.

Long Crowbar

A four-foot crowbar is handy when you're trying to move a tree over or lift it a bit to free a pinched saw blade.

Wedges

A heavy, thick metal wedge is ideal when you're trying to split very large pieces of wood. You hammer the wedge in with a sledgehammer until the wood finally gives way. Then you can usually cut around the broken edges with a maul to take care of the rest. A plastic wedge is used in felling, if you have a tree that is leaning one way and you're trying to get it to fall another. After you've got the felling cut started, you insert the wedge or wedges into the felling cut opposite the direction you want the tree to go and hammer it in.

Hydraulic Splitter

We're torn on the hydraulic splitter. It's a gasoline-powered machine, usually pulled behind a truck, that splits even the biggest and knottiest wood around. You can make quick work of a big tree and have a nice pile of evenly split firewood to show for it. So the time it saves is certainly a plus for a hobby farmer, especially if you need to get to the office or are only able to do farm work on the weekends. But farm work doesn't provide many opportunities for cardiovascular exercise. And chopping wood is one of the few types of work where you can sustain a steady, quick pace for an extended period of time. Unless you have a physical ailment that would prevent you from splitting wood safely or you have an entire forest of knotty walnut, we think you should split your wood manually.

Safety for Cutting Wood

✓ Wear safety gear—hard hat, protective goggles or a face shield, ear muffs or plugs, chaps, gloves, and heavy boots.

✓ Always let someone know when you're going out to cut and how long you expect to be gone. Take a cell phone with you if you can get reception.

✓ Never consume alcohol or any type of drug you wouldn't take while driving.

✓ Leave children and pets at home.

✓ Don't run a saw above shoulder height and never while standing on a ladder.

✓ Don't fell trees when it's windy.

✓ Beware of slippery or icy conditions.

✓ Don't cut a tree that's so large or is at such an awkward angle that it's beyond your expertise.

Felling

Felling a tree should be done with respect, both for the tree and for your own safety. As we suggested earlier, it's best to go out with someone with experience first to observe how it's done. Reading instructions in a book just doesn't do it justice. The high-pitched scream of the chain saw muffled through your earplugs, the smell of freshly ground wet wood in your nostrils, the heft of the chain saw bucking like a horse straining against its bridle, all create an otherworldly atmosphere that is best observed from a distance first.

As a hobby farmer, you might consider the time you have to deal with felling and splitting. A good way to shave off about half the work is to cut only trees that are thin enough that you can just section them up, dry them, and use them in your stove without having to lift a maul for splitting.

RIGHT: Don't attempt to cut a hung up tree like this one. It's best to wait for nature to bring it down.

Choose your tree wisely, looking to see if it can fall cleanly without being hung up or damaging other trees on the way down. If you can't see a clean line, then don't cut that tree or cut the other, smaller trees in the line first. Make sure that there are no obstructions around the tree that would trip you up or keep you from standing squarely while you work. Choose an escape route for when the tree begins to fall.

Once you're ready, make the initial cut parallel to the ground at about knee level and about a third of the way through the trunk on the side of the tree where you want it to fall. Use the bottom of the saw to make this cut, not the top of the blade. The handle on the saw is designed especially so you can hold it to make this parallel cut. Make the second cut at about a 45- to 60-degree angle down to the first cut. Make sure that the wedge is completely cleared from the cut. Now it's time to make the felling cut directly behind the wedge cut. This cut is also parallel to the ground and a few inches up from the parallel first cut. You want to make a hinge so the tree won't split awkwardly at the bottom when it falls. As you get close to the wedge on the front, you can feel the tree begin to shake. Don't cut all the way through to the wedge or your saw could be mangled and the tree could fall where you don't intend it. Cut the saw off and beat a hasty retreat. It's okay if you retreat too soon. Just go back and cut a little more until you feel the tree is falling and retreat again.

Barber Chairs and Widow-makers

Beware trees that lean too much. They can split violently under the weight of the tree directly back at you as you make the felling cut. This splitting and the resulting kick-out of part of the tree resembles an old-fashioned barber's chair. Many fatal and debilitating accidents have been the result. You should have an experienced feller show you how to treat leaning trees. But the technique is to make a much more shallow initial wedge cut of only a few inches deep. Then make smaller wedge cuts along either side of the main wedge cut before you make your felling cut. The side cuts will release some of the energy stored in the leaning tree and will keep it from spitting.

Hung-up Trees

Never try to cut a tree that has another hung up on it. Sometimes, it's just best to leave trees to let nature take its course. This is true even if you're the one that created the hang-up. It's just not worth the risk. It may take years for the trees to become free, but just think of all the happy moments you'll enjoy in your life during those years. Call a professional if you really have to get those trees down.

Limbing and Bucking Up

Once the tree is down, you can start limbing it. Start with the easy limbs first that are sticking up. Periodically clear away the cut limbs so you don't trip over them. You can turn the tree to get the limbs on the bottom, or sometimes it's best to leave one or two so they prop up the tree and make it easier to buck up. Limbing and bucking are the most common times when saw blades get stuck. Always be

Cutting Down a Tree

1. We chose this tree because the donkeys had eaten the bark and it was sure to die. Harold makes the initial cut parallel to the ground and knee high.
2. The second cut is made at a angle down to the first. Make sure to clear out the resulting wedge.
3. The felling cut is made behind the wedge, parallel to the ground and a few inches up from the first cut.
4. If done correctly, the tree will either break off freely like this one or stay attached without splitting at the hinge, which makes it easy to buck up.

LEFT: Use a timber jack to lift big logs off the ground so that they are easier to cut into sections.

thinking about how the weight is being distributed to the limbs and watch out for a lot of pressure on a limb. Cutting through a limb that's bowed under the weight of the tree will create a mini barber-chair effect and the limb could shoot out at you when you release the pressure by cutting it. Once you've limbed it and moved the debris away, then you can buck it up, which is cutting it into fireplace-length pieces. The shorter the pieces, the faster they dry, but also the more cuts you have to make. Don't just keep the big pieces you're going to split. Keep in mind that the smaller pieces from the branches are already the perfect diameter to dry and just throw into your stove without having to split them. And the really small stuff can be used as kindling. Try to use as much of every tree as you can. Many times the tree's butt is still slightly attached to the stump and you can prop up the tree off the ground a little ways down with a concrete block so you can cut without the ground causing your saw to get stuck. A timber jack is a very useful tool at this stage, especially for large-diameter trees. You just jack the tree off the ground and buck it up. Move the jack down the tree as you go and you'll make quick work of even the biggest trees.

Splitting

As we mentioned before, you'll want to use a maul to split your wood first. It's best to let your wood season for a while to make it easier to split. Some people will split wood right on the ground, but we find it easier to get a big, round, flat piece of wood as a chopping block. This brings the piece you're spitting up to a level that is easier and safer to strike. Use good work gloves as you'll easily get blisters without them. If you have a very big piece that won't split down the middle, try to knock a piece off the edge to get it started and then work your way around it. Really, really large pieces will need a wedge and sledgehammer. If you need your wood to dry out quickly or if you live in a very wet region like the Pacific Northwest, you'll want to further split your wood with an ax into very small pieces. It's more work, but the pieces will dry more quickly and burn hotter.

Wood-Splitting
Warning: Always wear protective eyewear when splitting wood.

1. Stand squarely with both of your hands at either end of the maul and your dominate hand under the head.
2. Raise the maul over your dominant shoulder and over your head.
3. As you reach the peak, slide your hand down from the head to meet your other hand at the bottom.
4. Bring the maul straight down, making sure you continue to follow through with your hands. Keeping your hands even or lower than the head as it comes down to contact the wood will ensure that you don't pull up, resulting in the sharp end pivoting on your hands and coming back at your shins.

Stacking

We stack in between T-posts driven into the ground and on top of old boards to keep the wood from being in contact with the ground. Leave the wood stacked to season before you split it. If you need your wood to dry quickly, you should split it and then stack it in a chimney—four to five pieces on a row and then four to five perpendicular on top of that and so on. This method also doesn't require the supports at either end, but takes a bit longer to stack. If you have two trees near each other, you can stack wood between them.

Ponds

A farm pond is one of the few gifts you can give yourself that will also benefit everything else on your farm. Not only can you fish, swim, and relax in it, but it will become a meeting ground for the local wildlife; it will serve as a way-station for migrating birds; and it will offer varied grasses and water for your farm animals. A pond situated nearby your house will also help to raise your water table and make your well water healthier. You can also use a pond to irrigate your garden or crops using a water pump, especially in times of drought. It can cost a good deal (ours was $10,000 for a half-acre pond), but the value it adds to your land is much more than what you will pay.

When deciding on a pond, you should consider exactly what you want to get out of it. If you want to have a fishing pond, it needs to be deep enough in your climate that the fish can escape the summer heat and shallow enough in some areas that they have spots to spawn. If you want a pond that attracts lots of wildlife, you need some long, shallow areas in which pond grasses and cattails will grow to give them cover and allow them to easily wade into it. The best ponds are designed by looking closely at the land surrounding you. Find the areas where the slopes of the land or streams converge naturally and envision how runoff or seasonal springs will feed it. The best ponds fit the landscape naturally.

The water level in most ponds is set by a pipe that runs under the dam and then comes up at a right angle to the desired water level. When the water rises above the pipe, it drains out under the dam and into a spillway or stream. But there are issues with pipes as they can become clogged with debris. We have to paddle out to ours in a boat every once in awhile to clean out the debris so that the pond won't overflow. Also, our pipe is slightly too small for the heaviest water flow times of the year, so it strains to keep the pond from overflowing at times. If you don't have

LEFT: In the foreground you can see the top of the pipe which collects overflow water in the pond and directs it under the dam. It needs to be cleaned out when leaves and debris clog the opening.

heavy flows, then a spillway is the preferred method. A spillway is just a rock-covered outlet that allows the overflow water to run around the dam. But spillway construction is a real science as they can easily erode and ruin your dam if you don't have it constructed properly to handle the flow.

It's best to have a pond professional consult and dig your pond. If you're building a pond of any real size, you need large earth moving equipment with an experienced driver and soil tests. Look for a pond builder or landscape architect and ask them what experience they've had building ponds. Find out how many they've built and what kind. Check out some of the ponds they've built so you can get an idea of how they do it and if you want them forever changing the landscape of your own farm. Ask questions and make sure to sign a contract that you study

carefully. You're going to be altering the terrain and water system of your farm and you should be completely comfortable with how it's done.

When building a pond, also make sure to sink posts into the bank before it fills up. These will be for your dock that you can build once you see where the pond level rises to. It's much easier to concrete these in at the beginning, before the pond fills up. A dock to relax on and fish from is a must. Plus, you'll want somewhere to dock your dinghy or flat-bottomed boat.

Now you've purchased your farm and have its "infrastructure" well underway, or at least have a good understanding of where it is heading. You've familiarized yourself with the basics of your farm, from the zoning ordinances to its energy sources to its water profile. You're now in a livable situation with a good deal of potential staring at you from outside your window. Time to begin imagining what you can grow, from fruit to flowers. Picture the creatures you might share your land with, from chickens to bees. Turn over in your mind the possible enterprises that you might create that will provide income to make your farm profitable or at least economically sustainable. A farm has endless possibilities and there's nothing more satisfying than pursuing them. Just don't try to pursue them all at once. Remember, you're a hobby farmer and you enjoy your lifestyle too much to gamble it away on risky farm ventures.

BELOW: Damsel flies have shed their shells for spring.

PART ②

Growing Things

CHAPTER 4

Where and How to Garden

As we contemplated our move to Virginia, one of the things we most looked forward to was starting our own vegetable garden. Throughout my childhood, my family always had a vegetable garden, although I mostly declined working in it. I still have vivid memories of my Uncle Kerry's garden—a lush and abundant vegetable crop lovingly and painstakingly fostered in a twenty-by-twenty-foot space in the back yard of his small home in Albuquerque, New Mexico. Even as a kid I wondered how he was able to produce such beautiful eggplant and copious amounts of delicious tomatoes in that small plot, while my family's veggie garden provided sparse and straggly lettuce, the required zucchini, and a few thirsty tomato plants in twice the amount of space. From these early observations

I learned the first essential truth of gardening: in order to have a successful garden, you must care for it constantly and diligently. Take the time to notice each plant and pay attention to its needs. What is your garden's number one need? Your time and attention.

My first real gardening experience occurred when I lived with my father for several months in the summer and fall of 1997. He had laid the groundwork by preparing some beds, and I dove right in. I had plenty of time as I'd just returned from a two-year teaching stint in Japan and wasn't quite ready to get back to work. I weeded, cleared out space, and turned over earth for a few more beds, and we were off. Strawberries, tomatoes, summer squash, winter squash, cucumbers, herbs, garlic—we were up every morning at 5:30 watering, checking for bugs, and weeding. And the plants responded. It was a wonderful garden, producing so much food we were giving away bags of produce to neighbors, my father's co-workers, and anyone who would take the extra off our hands. Of course, we also had to water in the evening, and check the plants during the day to make sure they hadn't succumbed to the crazy New Mexico dry summer heat. My father had been dabbling in gardening all his life, so he took care of much of the basic setup—soil amendments, plant spacing, and companion planting, and much more that I was to learn about as I began my first garden on my own in Virginia. In New Mexico it was all about the water and the sun—too much hot, blistering, drying sun, and never enough water. I was ill prepared for what Virginia had to offer in the way of moisture, soil, and bugs.—Audrey

Garden/Farm Journals and Record Keeping

As you begin planning your garden, keep track of all of your ideas, plans, expenses, and progress in a notebook or garden journal. Use this for all your farm enterprises as it's a good way to see how they all fit together and complement each other. This will become one of your most valuable resources. On the inside cover write down your hardiness zone and the average late and early frost dates for your area. Include a sketch of your garden and farm plan, attach useful articles or information, jot down websites, names, and questions. Most importantly, keep a detailed record of what you are growing. Make notes of what worked and what didn't, what garden pests and helpers you are noticing, the weather, things you want to try, and tips from other growers. This may seem a bit of overkill,

LEFT: Audrey's hand-painted garden flags help the cutting garden look less utilitarian.

and we confess that come the end of June, our notes become less detailed and are made less often, but the more you include in this journal, the better you will know your plants and animals, and the richer your experience will be.

Our Gardening Philosophy

Give some thought to your gardening philosophy. By this we mean how you will approach your relationship with the ecosystem you are about to create and with the environment surrounding you. Most people tend to think of gardening as an environmentally and people-friendly thing to do. It is seen as a productive endeavor from which the gardener gains quiet reflective time while producing something beautiful and worthwhile, and the earth from which we create our garden is painstakingly cared for and nourished. However, many of the aids to beautiful gardens and blemish-free produce are not friendly to the environment and can have long-lasting toxic effects on our land and water supply, as well as adversely affecting our own health and the health of the animals on your farm. Be aware that most of the studies and research on a product's environmental safety have been conducted by the companies and organizations that manufacture and/or profit from that product. Any chemically prepared commercial product should be carefully researched before use, even those that claim to be non-toxic and/or environmentally friendly.

Our garden philosophy is the same philosophy we have for the entire farm. We practice only organic and sustainable methods while encouraging healthy relationships between all the natural players on the farm — plants, animals and fungi. Each year we improve on these relationships by rotating crops and animals, for all of our mutual benefit. For instance, we now move our mobile chicken coops over the flower and vegetable beds when they're done for the season. The chickens eat the leftover vegetation, till up the soil, eat harmful grubs that might be trying to overwinter and fertilize the ground with their manure. In this way, the chickens obtain a healthy, varied diet that shows in the richness of their eggs that we eat and the soil is naturally enriched for planting again. A perfect loop of mutual benefit.

Soil Tests

It is universally recommended that you have your soil tested before you begin creating beds and planting your garden. Soil testing is a chemical analysis that determines the amount and availability of nutrients your soil already contains, and most test results include recommendations for amendments and/or fertilizers to improve your soil's fertility. There are home test kits that you can purchase online or at local garden centers; however, these generally test only pH—soil acidity. Simple pH test kits can be bought at any garden center. Contact your local cooperative extension office to obtain the necessary soil sample boxes and information forms for the more comprehensive soil quality tests. Most cooperative extension offices charge a small lab processing fee for soil testing services.

Cooperative Extension Agency

Before you get too far along in your farming and gardening plans you will want to investigate the website of your local cooperative extension agency. It's also a good idea to make an appointment for a farm visit from one of the extension agents.

The Cooperative Extension System is a nationwide, non-credit educational network that provides useful, practical, and research-based information on myriad topics related to rural living. Staffed by local extension agents whose specialties vary from fruit crops to cattle production, entomology to marketing, forestry to food safety and more, your local extension agency is your most reliable and accessible source of information and support. They are well versed in the particulars of your local growing conditions. They can suggest the best plants and methods for growing specific to your area. Find your local offices by going online to www.csrees.usda.gov/Extension/index.html where you will find a national map and can click on your state.

That said, let us visit Audrey's school of trial and error. You may want to start with only a couple of small raised beds, a sprinkling of seeds, and a few small veggie starts, and see what happens. This is what we did, and it works, up to a point. If you're starting with soil that's never been planted and only has grass growing on it (known as *new ground*), or if you are importing some soil from a reputable garden center to get started, then you may not need to test it before your first planting. Soil that hasn't been planted for many years or is mixed by a professional generally has a good balance. It's best to at least test the pH level to make sure it matches the growing range of the plants you will be growing in it.

Why should you have your soil tested? If you are going to the time and effort to create and plant a garden, putting in dozens of hours of physical labor and investing hundreds of dollars in building and planting supplies, you want your efforts to be fruitful—you want the crops you plant to grow and to be strong, healthy, and productive. Soil testing provides you with the information you need to give your soil exactly what it needs.

RIGHT: Tomato trellis made of cedar posts and electric fencing wire.

LEFT: Collect rain water everywhere you can. We use these old pickle barrels to collect water from the roof of our shed. We attached a screen to the top to keep out leaves and the inexpensive faucet and overflow tubing can all be bought at the hardware store.

Soil testing is an extremely valuable tool for those gardeners who have a particular crop in mind, whether it is blueberries, asparagus, or heirloom tomatoes. Each crop has different nutritional and pH needs. The better you meet these needs, the more robust and abundant your crop will be. Through soil analysis you can determine exactly what you need to add to your soil to provide the best nutrition and growing conditions for your particular crop. This also saves you money by allowing you to focus your efforts on specific needs.

The more years you garden, and the more care and time you put into that garden, the better your soil becomes and the better you are at figuring out what it needs. You want to know more, and you want to grow the healthiest, best plants around. But don't overwhelm yourself with soil science right away; all the chemical terms and various soil amendments can make your head spin before you gain a better understanding of the overall needs of your soil. It's an advanced gardening technique that you can learn over time. And so, at some point, you will want to have your soil tested. You may as well do it right from the start—and then do it again in a few years to see what, if anything, has changed. But a dump truck load of well-mixed soil and compost from a respected garden center is worth the money when you're first starting out and will save you time and worry about the exact composition of your soil until you're more comfortable in that area.

Garden Placement

Where will you plant your garden? Your plants will need lots of sun and access to plenty of water. Determine whether the area you are considering has good drainage. Take a look at what is growing nearby. If your garden plot is near a growth of poison ivy or poison oak, you may want to relocate. Also consider how much access you are offering to animal pests—deer, moles, voles, squirrels, groundhogs, (chickens!)—and how you may be able to thwart their efforts at making your garden their favorite snack spot. Tending and harvesting throughout the growing season is a daily task. Make sure your garden is in a spot conveniently near to where you will be cleaning your harvest, and readily accessible to quick picking for immediate eating.

Garden Placement Considerations

Sunlight—at least six to eight hours a day, preferably southern exposure. Eastern exposure is the next best thing, followed by western and lastly northern.

Water—access to source, ease of application

Soil—drainage, previous uses

Access—convenient to you, to farming/gardening tools, and machinery

We once planted three apple trees in a beautiful green sunny part of our field only to have them fall over in a heavy rain. Instead of moving them, we tried to prop them up and support them with twine. Not surprisingly, the roots actually rotted away from all the dampness. The spot was simply not able to drain off the heavy spring rains quickly enough for those trees.

Garden Layout

Once you have selected the location for your garden, you can begin thinking about its design. Do you want to set aside a special area of your garden just for herbs? Will your paths be structured for utility and economy (getting that wheelbarrow and other tools in and out easily), or do you envision a more meandering progression through your plants? This is your dream stage. Although it is not impossible to change the layout of your garden in later years, it is a lot of work. Thoughtful planning now will make your garden a convenient and fertile growing area for your plants, and a place you want to spend time. Get out some graph paper and sketch out ideas. Visit other people's gardens and ask what they like and dislike about their own designs. Think about what it is that you want to grow—what do you know you will eat? Some plants will need a permanent spot in your garden (perennials like asparagus) and these should be mapped out first; other plants must be rotated each

Define Your Farming Philosophy

Environmentally Sustainable Agriculture: Sustainable agriculture is one that produces abundant food without depleting the earth's resources or polluting its environment. It is farming that follows the principles of nature to develop systems for raising crops and livestock that are, like nature, self-sustaining. Sustainable agriculture is also the agriculture of social values, one whose success is indistinguishable from vibrant rural communities, rich lives for families on the farms, and wholesome food for everyone.

Organic agriculture:* "Organic agriculture is an ecological production management system that promotes and enhances biodiversity, biological cycles, and soil biological activity. It is based on minimal use of off-farm inputs and on management practices that restore, maintain, and enhance ecological harmony. The primary goal of organic agriculture is to optimize the health and productivity of interdependent communities of soil life, plants, animals, and people."

* (SDA National Organic Standards Board (NOSB) definition, April 1995, from website www.nal. usda.gov/afsic/pubs/ofp/ofp.shtml)

year (such as tomatoes and peppers), so you'll want to plan out beds of equal size and sun exposure to make this process easier. If you're planting in rows, it's best for them to run along a north/south line. That way, they sun hits the rows evenly as it passes overhead during the day and taller plants don't shade the shorter ones.

Your local library has several shelves of books devoted to gardening—browse through these and take notes. You will probably end up buying a couple of books that are the most useful to you. A selected bibliography of gardening books is included in our appendix, but it is by no means all encompassing.

As you browse through your pile of gardening books, you'll notice different methods of preparing beds for your plants. Small vegetable and herb gardens usually require raised beds with boards or other materials holding in the soil. But typical raised beds constructed with borders are not ideal for all crops. Long rows, grids, and trellising are optimal for certain plants. As you plan your garden, think about the size and placement of these gardening foundations.

After we had chosen our garden location—a mostly flat section of land with southern exposure about 100 feet from our back door—we fenced the area using cedar trees from our forested land and chicken wire. Our goal with fencing was to keep our dogs and chickens out of the garden, as well as keeping out the numerous wandering deer. Our garden fencing is very basic,

and it works for us. Other farmers have much more elaborate systems—necessary if you don't have great guard animals like we do!

Next we needed to rid the garden plot of the unwanted weed growth. Armed with gardening books, we figured out how to clear our designated garden plot of the insidious wire grass, also known as Bermuda grass, that covered the area. We wanted our garden to be as environmentally friendly as we could make it. We also wanted to get started right away. We quickly realized that in nature, the fastest way is rarely the most environmentally friendly.

Pesticides are pretty unpopular at this point in time, but herbicides can be just as toxic. They, too, can leach into the ground water supply and violate nearby streams and creeks. They also harm the organisms in your soil that make it healthy and productive. Many gardening books recommend the glyphosate herbicide Roundup™ as a safe, effective, and quick way to get rid of weeds. Many of the clerks at the local gardening supply stores also touted Roundup™ as a safe and easy to use product.

Our gardening philosophy includes the belief that part of gardening is learning to live and deal with pests that will not go away. We know that for many gardeners in our area, deer occupy this special place in their hearts and gardens. After more intensive research, we decided that chemically produced herbicides are not the right choice for our garden. For one thing, it's only a temporary solution and weeds will return with time anyway. We cover the wire grass with plastic, land-

scape paper, cardboard, straw, wood chips, pine needles, and cover crops We pull it up. We build our soil reinforcements every year in the battle. We rage and curse and fume at it. Gardening produces great pride and great humility. So, we battle the wire grass in a partially effective manner and move on to what we hope is a more rewarding pursuit, preparing the beds.

Bed Preparation

Our first summer on our farm was one of drought. Not that we noticed it as drought that first year. We knew because everyone around said it was a drought, a bad one. Restaurants in Charlottesville, twenty miles away, were using disposable cups, plates, and utensils in order to save water from dishwashing. To Audrey, a girl from the desert southwest, everything looked green and lush, and the humidity was a suffocating blanket of damp sweat-inducing misery. We had sketched out a plan featuring a series of raised beds built from 8x2-inch boards, alternating with a few dug out beds. Our plans quickly changed as we tried to dig out that first 4x4-foot bed. The project was much more like chiseling with a shovel than digging. The hard red clay soil of central Virginia was made even more impenetrable by the summer's prolonged drought. Stubbornly, we refused to give up on that bed. It became a symbol of getting our garden started. And so we painstakingly carved a perfectly square bed from that clay ground, and changed the rest of the garden plan to include only raised beds and raised rows. We ended up having soil mixed with compost delivered and used this for our first vegetable beds. On top of that, we mistakenly built all our raised beds with pressure-treated lumber. That is a big no-no, as pressure-treated lumber has arsenic in it that leaches into the soil and into your food. So we pulled them all up and rebuilt them with regular wood. Another lesson learned.

Remember, starting small is fine! You will make mistakes, and learning from small-scale mistakes is much less painful and expensive than learning from large-scale mistakes, especially since you most likely have an off-farm job that requires much of your attention and time. When we started our garden we knew we wanted strawberries—we love strawberries! We bought twenty-five bare root plants and planted sixteen of them into a beautiful, newly prepared 4x8-foot bed. The remaining plants went in between the newly planted asparagus roots in

their own beautiful, newly prepared bed. We dutifully pinched off the flowers the first year so that we would have gorgeous berries the following season. And we did. Luscious, red, abundant strawberries cascaded out of that bed. Their runners inserted themselves into the paths between our raised beds, and hopped over into adjacent beds. Those plants over in the asparagus bed didn't produce as much fruit (we learned that asparagus is a heavy feeder and doesn't like to share its bed with other plants), but they diligently sent out their runners, looking for more hospitable ground.

Our raised bed was not a good choice for the strawberry patch. On the positive side, we had hundreds of baby plants because of all of the runners. On the less positive side, we hadn't planned on room for a hundred or more strawberry plants. We ended up creating a long row for the strawberries right along our current garden fence. We dug up those plants in the raised bed, and transplanted as many runners as we could into that row. The following season we had an amazing strawberry crop, our freezer was full, we had more strawberry pies than we can remember, and certainly more strawberries than we could eat.

The drawback to the strawberries was the weeds. Little insidious weeds that twined themselves right around the crown of each plant, then grew in that rich, nicely watered soil. We spent many painstaking hours weeding that patch, and after three consecutive years, we finally took a break from growing strawberries. Granted, it was only a one-year break, but the vacation from this crop allowed us the time for more thoughtful planning.

Last season we planted seventy-five new bare root plants in a long row we had covered with black plastic. We had put down drip-irrigation before laying down the plastic, and inserted the plants into small holes we cut into the plastic. This helped tremendously with the weeds during the strawberries' growth season and we had an abundant and healthy crop of strawberries with much less trouble. While we hope to help you avoid much trial and error, don't be afraid to jump in and experiment. Much of the fun of gardening comes from learning your own lessons.

Since that first year's layout, our garden has grown and evolved, and we like to think we've grown and evolved along with it. A couple of the raised beds still occupy their original places. The only perennial vegetable we grow, asparagus, is still in its position near the back fence, but as we've refined our preferences and singled out things we don't want to deal with, we have adapted our garden to these changes. We stopped growing any kind of squash after the first two years for a few reasons: it sprawls and takes up quite a bit of space, we just don't eat very much of it, and we simply cannot seem to keep up with the squash bugs.

Compost

The key to successful gardening is healthy soil. And compost is the best way to improve soil and keep it healthy. Healthy soil will give you bigger yields, and stronger plants that are more resistant to pests and disease. It also discourages the worst kinds of weeds.

Chances are that you won't have perfect soil. Composting can help to improve all types of soil. First determine what you have and what you want, soil-wise. In New Mexico, we worked with a very sandy soil, good for drainage, but not so good for holding on to moisture and nutrients. In Virginia we have mostly a heavy clay soil—this makes drainage an issue, but not water retention. Good compost will improve soil, whether it has too much clay or too much sand.

Your goal is a rich, loamy soil that crumbles easily and readily absorbs moisture. The best amendment for any soil is your own homemade compost. Composting is using the natural process of decay to change organic wastes (leaves, grass clippings, kitchen scraps, manure, etc.) into a nutrient-rich humus-like material. Microorganisms and invertebrates (bacteria, fungi, microbes, earthworms, and insects), which are already present in the materials to be composted, break down this organic waste, and composting systems are based on providing optimal conditions in which these little guys can do their work. The best part about using compost as your main soil amendment is that you don't have to be a chemist. Good compost is well balanced and takes care of the soil's needs on its own.

All organic material will eventually decompose. "Cold" composting is what happens in nature, particularly in shaded areas, as leaves and other organic debris slowly turn into a rich, humus loam over a period of several years. That's why you find such dark and rich soil in forested areas.

RIGHT: After you've constructed your beds, instead of tilling use a heavy-duty broad fork to loosen soil and mix in amendments. It does less damage to the soil structure and kills fewer worms than a tiller. It's good exercise too.

Hot Composting

We can hasten the speed at which our compost pile decomposes, creating "hot" composting. It's called hot because the soil is heated up to 135 degrees or more by the process of decay. In order to make this work, you'll want to monitor:

- the ratio of carbon to nitrogen in your pile (see below)
- size of your pile and the amount of surface area exposed
- aeration, or oxygen in the pile
- moisture of the pile
- temperatures reached in the compost pile
- outside temperatures

The microorganisms need both carbon (C) and nitrogen (N) food sources. An optimum C:N ratio for speedy composting is 30:1. Carbon provides the energy for the microbes to help break down the compost materials. Carbon sources are called "browns" because they are generally brownish or darker in color. Browns are generally dry and slow to decompose. Nitrogen provides the elements of protein that the organisms need to grow and reproduce. Nitrogen sources are called "greens," although not all greens are green in color. They arc usually materials high in moisture and are quick to decompose.

Browns: straw, leaves, wood chips, sawdust
Greens: vegetable scraps, grass clippings, coffee grounds, manure

Hot composting is all about providing the best possible working environment for the microorganisms in your compost pile, so it is important to pay attention to the size of the materials you are adding to your pile. Smaller particles of material have more surface area for microbial activity, and are also easier to mix as you turn your pile. Generally, a particle added to your pile should be no more than an inch in diameter, unless you're not in a hurry for the pile to finish. Chop or shred any larger particles before adding them to your compost pile. We like to layer straw (brown material) between heavy green material like kitchen scraps to keep the pile aerated and to avoid it becoming compacted.

For efficient hot composting the pile must be big enough to hold heat. A hot pile decays much faster than a cold pile. A pile of about one cubic yard (3x3x3 feet) is big enough for year-round composting, even in cold-winter areas. But the bigger the pile, the better. Big piles heat up quickly. Of course, the temperature outside the pile does affect the temperature inside the pile. Your composting season closely matches your growing season. Your compost pile will decompose much more slowly in the coldest winter months. The optimum temperature in an active compost pile is 135 to 140 degrees F. At this temperature most weed seeds and disease organisms are killed off. A properly built pile will reach optimum temperature after about a week. At this point the pile should be turned or stirred so that material from the sides is moved to the center and any compacted materials are loosened. Also check the moisture of the compost. Microbes function best when the compost heap has many air passages and is about as moist as a wrung-out sponge. You can hasten the composting process by turning your pile often. As materials decompose, the pile heats up and should also shrink, eventually becoming about half its original size. Once the hot phase is complete, lower-temperature microorganisms, fungi, worms, insects, and other invertebrates complete the decay process. Depending on the factors discussed above, your compost should be ready to use in about three months.

There are many ways to go about starting and building your compost pile, and there are many philosophies regarding the "best" way to build a compost pile, from the lasagna layering method to the throw-it-all-in-a-pile-and-just-leave-it method. Adding kitchen scraps (vegetable peelings, coffee grounds, etc.) and yard waste (such as grass clippings and leaves) to your compost pile has the added benefit

of being an environmentally sound method of recycling this waste. Remember not to add any meat or dairy products to your compost, and never add cat or dog feces to the pile! Both have dangerous pathogens that may not be killed during the composting process. Many excellent books on building your own compost pile are available, and it is worth the time to browse through a few, especially if this will be your first compost pile. Most gardening books also contain a section dealing with compost and how to make it. You can also hop on the Internet and find hundreds of sites devoted to or containing information about composting.

No matter how you decide to go about it, having a compost pile is an essential part of your garden endeavor. Even if you just pile up leaves and grass clippings and cover them with an old tarp, you will be creating compost, and compost is your garden's best friend.

One thing you'll figure out early on is that it's virtually impossible to create enough compost for your garden from your kitchen scraps alone or by using one of the many compost tumblers on the market. The best single source of compost material is vegetarian animal manure. If you don't have your own manure producers, most rural areas abound in farmers who will either give manure away to those who load and haul it themselves, or are selling it. Fresh manure contains high levels of nitrogen that will burn your plants, and pathogens that can contaminate your vegetables, so don't be tempted to toss it directly into the garden. Make sure to compost it before mixing it into your soil. Horse, cow, donkey, sheep, goat, rabbit, and chicken manures are all considered hot manures and must be composted before adding to your garden soil. In addition, horse, cow, and donkey manure can also contain large amounts of weed seeds, which are mostly eliminated during the hot composting process. Bottom line: compost!

Because we own llamas we have ready source of cold manure. Llama pellets can be worked directly into the garden soil without burning your plants. Adding them on top of the soil is less effective since the pellets tend to dry out rather than break down into the soil. We add them to our compost pile, or mix them with the donkey manure and allow them to break down and age. Another way to break down llama manure is to form a pile and moisten it until damp, not soggy, and completely cover the pile with plastic. In about six weeks your pellets will turn into a substance the consistency of peat moss that is easily incorporated into your garden soil.

As you are planning your garden's layout, consider where you will put your compost piles. If you will be using the vegetable scraps from your kitchen, you'll want to think about the ease of carting those scraps out to your pile. If you'll be bringing in manure, you'll want to unload it as close to the pile as possible, so vehicle access is something to consider. Also keep in mind that in prolonged periods of dry weather you will need to add water to your pile to keep it moist. By locating your compost pile in a convenient spot, you will save yourself labor in the future.

We have three compost piles going at different stages at all times so that we can have ready compost whenever we need it and we always have a young pile to add our kitchen scraps to so as not to slow down the finishing of the other piles. We also maintain a manure pile and a mulch pile to use when creating a new compost pile. We keep a small plastic container in our kitchen for food scraps. When that's full, we move it to a five-gallon bucket outside the back door with a tight lid. When that fills up, we cart it out to the compost pile and add it to the youngest pile with about an 8-inch layer of straw. Once this pile heats up (after about a week), we mix that all together, make sure it's moist enough (adding water if needed) and start again with empty compost buckets.

When is the compost pile finished? Scoop some up, smell it, and find out. The finished compost should smell like fresh earth—it should not have any unpleasant or sour odors of decay—and should look dark and crumbly. You may still see some materials that haven't broken down fully, like nut shells, but as long as the pile smells fresh, then it's ready to use. Do not use compost before it's time or you might injure your plants or introduce weeds or disease into your garden.

BELOW: Our three compost piles, separated by old fencing and old straw bales. The pile on the right was just created and is too hot to use in the garden. The pile in the middle is almost there, but needs a few more weeks. The pile on the left is just right and ready to be worked into the garden.

Cover Crops

Because we started small (our first goal was a kitchen garden), several raised beds in an approximately 1,500-square-foot space, and were eager to have vegetables as soon as possible, we did not even think about cover crops. In our minds, cover crops were associated with large-scale commercial production farming. Cover crops have traditionally been grown mainly to prevent soil erosion by wind and water. However, as our garden expanded, we had more area of soil to attend to—more weeding, more soil amending, generally just more work. This is when our reading and conversations with other farmers led us to cover crops. Cover crops can be grown for "green manure," or as a "living mulch."

Green manures are crops that are turned back into the soil just before they are about to seed, which then adds their nutrients back into the soil. To incorporate these plants into the soil, most people use a tiller. This method works, but it damages the soil and mixes its layers. The best way to incorporate green manure is to cut it with either a trimmer or lawnmower and then work it in with a broad fork, and let it decompose into the soil for at least six weeks before planting anything else. This will ensure the bacteria has time to enrich the soil. A cover crop can be grown in the fall, mixed in before winter, and left under a layer of straw. Or you can grow a hardy crop through the winter and incorporate it in the spring. Living mulches are crops that are grown to help suppress and crowd out weeds and prevent soil erosion. These crops need to be kept mowed to prevent them from going to seed and thus taking over your garden. Or you can choose a versatile cover crop that both crowds out weeds well and is rich in nutrients.

Weed Control and Soil-Building: They Go Hand-in-Hand

Weeds are a constant battle, but not one that needs to deter you from gardening. The single best thing you can do for weed control is to develop healthy soil, full of worms, fungi, bacteria, micro-organisms, and microbes. Weeds grow best in unhealthy soil. But it will take many years to develop your soil to a level that discourages weeds. So you'll need to use other tools along the way that also help your soil's health.

In Virginia, the worst weed offender is wire grass, a nasty plant that travels underground via runners and wraps itself around the roots of other plants. If you don't kill the entire plant, the remaining runners will live on, like an alien cancer. But wherever you are, you'll have your own demon to battle. We do not advocate herbicides as they kill everything in the soil and above it, including worms, microbes, and bees. But we also know that there might be severe cases of infestation of poison ivy or some other noxious weed that require tougher methods.

Here are several weed control and soil-building methods that we've found most useful.

Build Your Soil and Fight the Weeds

Compost—Keep adding compost to your beds to build up their health and immunity to weeds. Healthy soil discourages weed growth, so keep making and distributing your compost.

Cover Crops—Two successive plantings of winter rye or buckwheat will crowd out even the worst weed offenders like wire grass, while adding nutrients and organic matter to your soil. There are many different types of cover crops, including clover and legumes. Rye and clover are the best for weed control, but many others offer different nutritional benefits, so read your seed catalog carefully to pick the one that fits your needs.

Shovel them out—Dig deep with a shovel and pull the weeds out by the roots. This will help loosen your soil at the same time.

Lay cardboard or at least five sheets of newspaper (we get big loads from our local recycling drop-off) in your pathways and then cover them with thick layers of wood chips, mulch, or straw. This will effectively stop any weed. But even the smallest gap in the cardboard or newspaper will allow a weed to take root. When this materials, breaks down, it becomes valuable organic matter that you can then shovel onto your beds. Then re-mulch your pathways and start over. In this way, you are fighting weeds while building healthy soil in your pathways to use on your beds.

BELOW: Annual rye grass is growing in these two beds. After the sunflowers were cut, we parked the chickens over the beds for a few weeks. Then we tilled the pathways, shoveled them onto the beds, and planted the cover crop. The chickens are now working on the two beds next to them.

Lay cardboard or black plastic on your bed for several months during the growing season. Then take it off and cover the bed with a heavy layer of straw for the winter (pull the straw back before planting). You'll have a weed-free bed in the winter and the lower layer of straw will have broken down to provide more organic materials to work into the soil.

Mechanized Tilling—This is the least desirable option after herbicides. Tilling mixes the soil layers and kills worms and valuable micro-organisms in the soil. You can offset some of the damage by tilling in compost or cover crops, but not all of it. We've been able to avoid tilling in our vegetable garden for several years as we've developed the soil carefully. But we've not completely moved away from it in the field where we grow our flowers; mostly the field is too big to realistically avoid tilling. As hobby farmers, we have to continually judge the amount of time any chore will take against the demands of Michael's off-farm job. Working our flower bed soil into shape will take several more years that include mechanized tilling, but we're slowly incorporating no-till gardening one row at a time. Hobby farmers don't have time to be soil purists, but the better you treat your soil, the less you'll have to work to grow healthy plants in the long run.

Pests and Disease

There are good bugs and bad bugs. It's to your garden's benefit that you know the difference. Buy a bug book and spend some time identifying bugs in your garden. Bugs like the praying mantis, the aptly-named assassin bug, and lady bugs are all good bugs as they prey on others that do damage to your plants. We practice only organic pest control. So what can you do if you don't use pesticides and you have a pest problem?

RIGHT: The only proven way to keep deer out of your garden is a good fence. No deer yet has penetrated this 7ft., woven-wire fence with two electric stands along the top protecting the garden at Broadhead Mountain Farm.

Organic Pest Control

Crop rotation—Bugs winter down around plants that they like to eat. Move those plants and the next season those bugs will have trouble finding them.

Chickens—Let your chickens loose in your garden at the end of your growing season. They will dig out many grubs and other bugs that are trying to over winter around your plants.

Floating row covers—For whatever reason, bugs don't like to operate under row covers. They also have a harder time getting to plants because they can't fly to them. Row covers early in the season keep out bugs that want to lay their eggs in spring.

Cleanliness—Clean out all your dead plant material at the end of the season (or when it's dead) and either burn it or move it very far from your garden to let it compost for several years.

Hand Picking—You can stay ahead of a problem by handpicking the bugs off into a cup of soapy water. Be vigilant and try to catch problems early. Find the eggs and kill them too.

Dish Soap—Many bugs, like aphids, will die from just a spraying of soapy water. Mix common dish soap with water in a spray bottle and go to work.

Cats—for larger pests like mice, moles, and voles, cats are the only effective remedy. Make sure you have at least one cat that sleeps outside at night as this is when the rodents are most active.

Organic Pesticides—While labeled organic, there haven't not been enough studies to know exactly how and what effects organic pesticides might have on humans. They are also very expensive. So use them as a last resort and wear a mask, eye protection, and protective clothing when spraying them on.

Diseases are less easy to deal with. Practice the good gardening techniques that we describe in this book like crop rotation, soil-building and composting, the cleaning out of dead plant material and use of cover crops, and you'll prevent many diseases. Use books or the Internet to identify your problem and deal with it immediately. If disease is only present on a plant or two out of many, sometimes it's best to just remove those plants or at least the parts of the plants that are affected. As with pesticides, there are organic applications for specific diseases that are expensive, but sometimes that's the only remedy. Each year, there will be some crop of ours that falls victim to disease. This year, late blight attacked our tomatoes. There's always next year.

Shiitake
mushrooms
growing on
oak logs.

CHAPTER 5

The Food Garden

What you decide to plant in your garden will depend first on your soil, water, and light conditions and your climate/hardiness zone, and second on your idea of how your garden fits into your life—what you like and want to grow, and how much time and trouble you are willing to take to make it happen.

We plant our vegetables almost exclusively in raised beds. Many vegetable plants do quite well when direct seeded in the garden, especially root vegetables such as carrots and beets. Greens (spinach, lettuce, kale, etc.), cucumbers, and squash also germinate and grow well when direct seeded. Other vegetable crops need to be planted as seedlings or "starts." These are available from local farmers at your farmer's market, local greenhouses or nurseries, or at the garden

Plant Hardiness Zones

There are eleven USDA Plant Hardiness Zones in the United States and southern Canada. The USDA planting zones are regions defined by a 10-degree-Fahrenheit difference in the average annual minimum temperature. The higher the numbers, the warmer the temperatures for gardening in those planting zones. It is standard practice for seed dealers and nurseries to label their products according to their USDA planting zones, thus indicating the planting zones in which you'll be most successful at growing those particular plants.

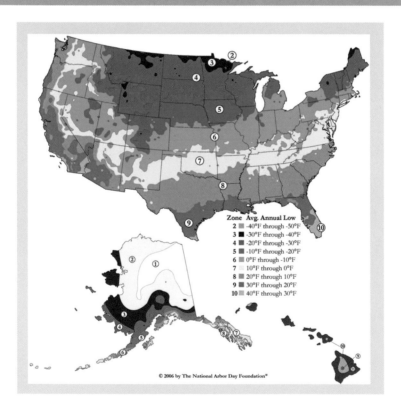

Zone	Avg. Annual Low
2	-40°F through -50°F
3	-30°F through -40°F
4	-20°F through -30°F
5	-10°F through -20°F
6	0°F through -10°F
7	10°F through 0°F
8	20°F through 10°F
9	30°F through 20°F
10	40°F through 30°F

© 2006 by The National Arbor Day Foundation®

centers of most home improvement stores. Because they take a while to grow, and they prefer warm soil and temperatures, tomatoes and peppers are good examples of these crops. If you have the inclination and the space, you may want to start some of your vegetables from seed on heat mats and under grow lights inside to give you a head start on the growing season. For many years we bought tomato seedlings

Starting Seeds

If you decide to give starting your own seeds a try, it doesn't have to be expensive or complicated. These are the basic materials we use and where to get them:

- Shelving—A wire rack (purchase at home improvement store, or make your own, such as board and cinderblock)
- Fluorescent lights
- Heat mats (purchase from gardening supply or seed catalog company)
- Seed starting medium (available from local garden centers)
- Seedling trays and pots (these can be recycled containers such as large spinach packages and yogurt cups)
- Fertilizer (we use organic fish emulsion ordered online)
- Automatic timer (purchase from gardening supply or seed catalog company)

to transplant into our garden. A couple of years ago, when we started growing cut flowers from seed inside, we decided to try to grow our own tomatoes from seed. Not only did this give us a greater selection of tomato varieties from which to choose, but it also allowed us to control exactly what contributed to our seedling's growth. The plants wound up being stronger and more resistant to disease. Now we grow all of our own starts and take the extras to the farmer's market.

Make a list of the vegetables, herbs, and berries you would like to grow. Look through seed catalogs and take note of different seed varieties, prices, and plant requirements. Most seed catalogs offer quite thorough explanations of the growing needs of their seeds. They should include the hardiness zone for the plant, its water, light, soil, and nutrient requirements, its resistance to certain diseases and pests, and its growth habit (how much and what kind of space it will take up in your garden).

You don't have to order seeds from a catalog, but by browsing through the offerings of a few different companies you will get a good sense of what is available, what will meet your growing qualifications, and what you want to look for as you shop. You can purchase seeds as well as transplants at your local gardening venue; this can be more economical by avoiding shipping fees if you are starting small. And in most

RIGHT: You can start hundreds of plants with just a simple set-up like this one in the basement at Broadhead Mountain Farm.

cases, locally grown seeds are more adapted to your local climate and have a better chance of success.

Once again we found the agricultural cooperative extension agency's website to be most helpful when we began our vegetable gardening in Virginia. We found a comprehensive and detailed document of vegetables recommended for Virginia complete with planting and harvesting dates. Getting your plants in the ground or seeded at the right time can be very important for certain crops, and can help you make the best use of your space if you want to plant some cold weather crops in the fall.

Popular Crops and Best Practices

There are many terrific books on growing organic vegetables. In this book, we'll discuss only our favorite plants and those that sell well at the farmer's market. Also, each plant has standard practices for growing and we won't cover all of those either. But what we will do is give you that extra helpful advice, or what we call "best practices" that we've found from our own experimentation and that of our farmer friends to create the best possible environment for growing each vegetable we discuss.

Tomatoes and Peppers

Starting—Plant your own seed starts in late winter in northern zones and early winter in warmer climates. You'll have to play with the timing each year to get it just right for your specific location. Harden them off for a week or two before planting outside by moving them outside to a cold frame or for a few hours each day on a garden rack.

BELOW: Heavy-duty concrete meshing like this keeps your plants upright and allows them room to spread.

BELOW: Plants like cucumbers and peas like to be trellised. Here's a simple design made from cedar posts and leftover television cable wire.

When to plant—Plant after the last frost. Don't rush it, as tomatoes and peppers won't grow until the soil is sufficiently warm anyway and you run the risk of killing them in a late frost. The only difference here is that peppers enjoy a cold snap, so it's best to move those plants out into a cold frame during the time that night temperatures are still dropping to the low 50s. Move them back into the house until the last frost is definitely past.

Where to plant—Full sun in raised beds or raised bed rows.

Best practices—We've tried every conceivable method, from those flimsy tomato cages (which don't work) to building a trellis and training the tomato vines up vertical lines every eight inches or so. This method works pretty well, but it makes it difficult to rotate your crops and you need to prune the suckers (non-fruit-bearing side branches). By far, the best method we've found is that used by Wally Parks at Broadhead Mountain Farm. He uses big four-foot-wide concrete meshing—the kind that looks like heavy-duty woven wire fencing with squares that they lay down when putting in a concrete slab. He rolls it into cylinders using five-foot lengths and secures them with fencing wire. They are heavy and strong enough to just set on the ground around your tomato plants. They easily hold all the weight of the big plants, and as they grow, the setup just becomes more stable. Use shorter cages of the same material for the peppers. There's no need to prune the plants at all. This allows maximum sideways spreading for the plant, adequate air circulation, and the most exposure to the sun.

Asparagus

Starting—Order asparagus as one-year-old plants from a seed catalog. Starting from seeds is for advanced asparagus aficionados.

When to plant—Plant them in the fall in most climates or early spring if you are in the northernmost part of the country. It takes two years to really start getting a good crop and three years to get a full crop.

Where to plant—In a raised bed in the back of your garden, dig deep as they have deep roots and mix in a large amount of compost. But while you dig deep to loosen the soil, don't plant them too deep as they can easily rot. Because they are an early-bearing perennial crop, there won't be a lot of other plants to shade them. Also, asparagus grows quite large into big, wispy green plants after you're done harvesting. So keep

them in the back of the garden, as you'll not want them to shade other plants later in the season.

Best practices—As a perennial, asparagus benefits from relationships with fungi in the soil. So you might consider bordering your raised bed using mushroom logs (discussed later in this chapter) or spent mushroom logs. The asparagus doesn't need a lot of sun and later in the season it will provide shade for the mushrooms. Keep weeds far away by laying cardboard in the pathways around it and covering with wood chips. Test the soil before planting. Asparagus likes a 6.5 to 7.5 pH range. Add amendments as needed. Don't harvest the first year, harvest only a couple of spears per plant the second year, and then you can harvest the best spears from then on. Nothing sells at the farmer's market like asparagus. It's impossible to grow enough.

Potatoes

Starting—Order seed potatoes from a seed catalog.

When to plant—Potatoes can be the first thing you plant in the spring.

Where to plant—In a raised bed or row with deep soil. Full sun is preferred, but partial shade works too. Potatoes like a slightly more acid soil than many other vegetables, between 5.0 and 6.5pH. Plant the seed potatoes in a shallow trench about a foot apart.

Best practices—Instead of hilling the potatoes with dirt, the best method we've found is to use straw mulch. The first layer you use should be shredded, so that it packs down for several inches. As the plants grow, periodically add straw around them to keep the plants upright. They can grow to be three feet tall or so and you'll end up with a large mound of straw that retains lots of water, which the potatoes love. You'll need to keep the straw from falling out the sides of the bed by lining it with chicken wire or even just branches or old wood. Harvest new potatoes after a couple of months (when you see flowers form) simply by moving aside the straw and digging down with your hand to pull them out. Once the plant has died off later in the summer, you can leave the potatoes in the dirt and they will stay nicely preserved as long as it's not really wet. You can then harvest as needed by digging down around the edges of your bed and lifting the soil up with a shovel or broad fork. When it's all done, you also have partially composted straw to use around your garden or in your compost pile.

RIGHT: Potatoes grown vertically. As they grow, support them with straw.

Garlic

Starting—It takes years to build up a garlic crop of your own. Each year, you should save the best heads for planting. You can also order more from seed catalogs. We've not had luck just buying from a grocery store and planting them. But a good strategy is to buy the best heads for sale from an organic grower at your local farmer's market. This is an expensive route, but at least you'll know that the variety they are growing does well in your area. You should plant in the fall a month or so before the first frost. Break the cloves apart and plant them about two inches deep and six inches apart with the point up.

Where to grow—Prepare a raised bed with loose, compost-rich soil. You don't want heavy soil that retains too much water. Mix in some sand, if necessary. The pH should be around 6.5. Cover with a heavy layer of straw through the winter, but pull it back after the first frost.

Best practices—When the leaved stalk puts out a central flower, or scape, it will resemble a pointy bulb at the top of the long stem. This will curl around itself. When it begins to uncurl, cut the scape at the stem right above the top leaf. (Scapes are quite tasty, chopped and sautéed like garlic, and sell very well at the market. We sell them for about a quarter each, which helps us recover some of the cost we've put into each bulb.) Stop watering once the scapes have come in, or the bulbs can rot in the soil. Once the bottom leaves begin to turn brown, but there are still five green leaves, it's time to harvest. Don't just pull them up; dig around the edges and pop the bulbs out so you don't damage them. Hang them in groups of about half a dozen in a hot, dry place out of direct sunlight until the papery protective covers are well developed. Then store them in a cool, dry place. Don't forget to set aside the best heads for planting the next year.

Greens

Greens include any leafy green vegetable harvested to eat its leaves i.e., spinach, kale, or broccoli rabe.

Starting—Greens are best grown from seed. They are really easy and require very little soil depth or nutrients. No need to buy starts as greens grow quickly in the proper environment. Plant in the fall directly into the soil. Depending on where you live, you may be able to grow them all the way through the winter. Heat is the enemy of greens, not cold. You'll be surprised how hardy they can be in the coldest months while you're enjoying a fresh green salad with your hot soup.

RIGHT: Greens growing in a cold frame.

Where to plant—Plant almost anywhere in good soil. The roots do not go very deep, so you can just plant into a few inches of compost. No need to mulch as they typically crowd out other weeds. And since they are growing in the cold weather months, they don't have much competition to begin with.

Best practices—Plant in a cold frame or an unheated greenhouse. In the cold frame, cover the ground with cardboard. Pile up six or so inches of soil and compost (or even exclusively compost). This creates the perfect environment for greens and you should enjoy harvesting them most of the winter. It's the young, small leaves (or baby greens) that are generally the tastiest.

Crop Rotation

Crop rotation is important for soil health and pest and disease control. Plants in the same family generally require the same nutrients, and are subject to the same pests and diseases. In terms of soil health, each crop takes up certain nutrients from the soil. If a crop is left in one spot year after year, that soil becomes deficient in the nutrient the crop has been taking up and the crop cannot thrive. The same is true for pests. Many pests overwinter in the soil. When they hatch or mature in warm weather, they're looking for their favorite food that was in this spot last year. If you've rotated that crop to another bed, the bug doesn't know where to look for it. If it has to waste energy looking for its favorite crop, it is less likely to survive or procreate. Below is a chart of plant families. These are the groups of plants you'll want to rotate around your garden.

Being aware of which plants belong to which family helps you make wise decisions when planning where to plant your crops each year. When you are trying to eradicate soil-borne diseases, allowing at least two to three years between similar crops is necessary. Crop rotation can also help to improve your soil's consistency by planting deep-rooted vegetables where shallow-rooted plants previously grew. Most gardening books contain plenty of good information about crop rotation and offer sample rotation plans and garden maps. It doesn't have to be complicated, just pay attention to what you plant and where you plant it and record that in your garden journal.

LEFT: An heirloom tomato ripening on the vine.

Family Name	Crops
Apiaceae	carrots, parsnips, celery, dill, chervil, cilantro, parsley, caraway, fennel
Asteraceae also known as Compositae	lettuce, endive, escarole, radicchio, dandelion, Jerusalem artichoke, artichoke, safflower, chicory, tarragon, chamomile, echinacea, sunflowers
Brassicaceae also known as Cruciferae	cabbage, cauliflower, broccoli, horseradish, kohlrabi, kale, Brussels sprouts, turnips, Chinese cabbage, radish, rapeseed, mustard, collards, watercress, pak choi, bok choi, rutabaga
Chenopodiaceae	spinach, beets, chard, sugar beets
Cucurbitaceae	cucumber, melons, watermelon, summer squash, pumpkin, gourds, winter squash
Ericaceae	blueberries, cranberries
Fabaceae also known as Luguminosae	beans, peas, lentils, peanut, soybean, edamame, garbanzo bean, fava beans, hairy vetch, vetches, alfalfa, clovers, cowpea
Lamiaceae	lavender, basil, marjoram, oregano, rosemary, sage, thyme, mints, catnip
Liliaceae	asparagus, onions, leeks, chives, garlic, shallot
Poaceae	corn, wheat, barley, oats, sorghum, rice, millet, rye, ryegrass, sorghum, sudan grass, fescue, timothy
Polygonaceae	buckwheat, rhubarb
Rosaceae	apples, peaches, apricots, nectarines, plums, strawberries, blackberries, raspberries, pears, cherries
Solanaceae	peppers (bell and chile), tomatoes, potatoes, eggplant, tobacco, tomatillo

Over the years, we've figured out how to make our garden and gardening more enjoyable, efficient, and productive:

The three things that made the biggest difference were raised beds, mulched, wide paths and drip irrigation.

We also learned that tomato trellises and cages, landscape fabric, cold frames, row covers and, hoop houses and greenhouses improved that plants quality and quantity.

Growing Berries

Berries are among the most enjoyable, sustainable, and profitable of any plant you can grow. They generally fruit early in the season before the dry weather sets in, before the bugs come out, and before there is a big selection to choose from at the farmer's market. Because most are perennial (strawberries not included), you only need to make one investment to buy the plants that will bring bigger fruits and larger harvests for many years to come. And they don't require too much in the way of nutrients or upkeep. But like every plant, the more attention you pay and care you give them, the better your results will be. The only drawback, if you can even call it that, is that it's very labor intensive and time-consuming to pick a large amount of berries. But you only have to do it for a few weeks during a season and that labor will fetch you a good $5 to $7 a pint at market, so that's no excuse not to get started.

Strawberries

As we mentioned before, strawberries are one of our favorite fruits. Strawberry plants can be purchased at your local garden center or farmer's market, or more economically as bare root plants from various seed and plant companies. A great number of strawberry varieties are available to today's growers. These varieties fall into three general classifications by when and how they bear fruit.

June-bearing, or summer-bearing, strawberries produce one large crop during the season. This means you will get a lot of strawberries at one time, so plan on an intensive harvesting and preserving period of about three weeks. Even within this class, however, are early-, mid-, and late-season bearers so you can best plan for your climate and best time for harvest. In general, strawberries ripen about thirty days after blooming.

Everbearing varieties produce a crop in spring, then produce a second, smaller crop in late summer. Some varieties produce several smaller crops approximately every six weeks. The berries of the secondary harvests tend to be smaller and the overall yield of the season is about the same as for June-bearing plants.

Day-neutral strawberries have been bred to fruit throughout the growing season. Trimming back the runners on these plants, and consistent picking of ripe berries, will keep these varieties performing well.

Investigate the strawberry section of your various seed and plant catalogs. Breeders are continually coming up with new varieties to suit specific regions, climates, and growing preferences. We tend to prefer the June-bearing varieties because we can harvest and preserve the crop before the rest of our garden demands our full attention. Also, in Virginia the heat and the bugs are intense during the height of the summer, so we like to grow the early bearing fruits in order to avoid the drought and pests that will come later.

Strawberry Growing Needs

- Slightly acidic, fertile, loamy soil
- Full sun
- About one inch of water a week while ripening

Strawberry Growing Methods

Strawberries take two years to produce a harvest. Most varieties are planted in the spring. Runners (horizontal stems with new plants) extend from the "mother plant" and take root nearby, forming new plants. The plants will produce blossoms (which if left will turn into fruit) that should be pinched off this first year to encourage strong root systems and plant growth. The following year the plants will produce an abundant crop of large, delicious berries.

There are many ways to plant strawberries. Here are three of the most common:

Row system—Plant strawberries in rows twelve inches apart and allow the plants to spread runners to cover the bare ground to the point at which you have about one plant every three inches. Pinch back the blossoms the first year and harvest the second year. Once you harvest the crop, plow the plants under and repeat this process in the spring. This method will provide a crop every other year unless you maintain two staggered plantings.

Hill system—Prepare two parallel raised rows twelve inches apart and cover with black plastic or landscape fabric (you'll need to run drip irrigation under the plastic or they won't get enough water). Cut holes every ten to twelve inches and plant a strawberry plant in each hole. The plastic or landscape fabric will prevent weeds from taking over your strawberry patch, saving you time, energy, and frustration later. Cut back the runners as they appear. As usual, pinch off the blossoms the first year, but after the first harvest berries should be harvested annually. Fruits will be larger with this method because the plants are not investing energy into producing runners; however, the plants will tend to produce abundant harvests for only about four years.

Refurbishing an old patch—If you've got a row of strawberries that has not been tended, you can reinvigorate it. Mow or weed-whack the middle of the row and the side on which you don't want berries to grow. Till the mowed section under and till it again several times (strawberries are tough and will survive one tilling). Then, as the berries you've left growing send out runners, move them all to one side (whichever way to you want them growing) and weight the runners down with a small stone or landscape pin. The runners will root and become new, vigorous plants. It's the new plants on which the best berries grow. In this way, you can keep your strawberry patch on the move across your garden (that is, if you have room for it to be moving each year).

Raspberries

Red, yellow, black, or purple, single crop or everbearing—these are some of the choices you will have to make if you decide to grow raspberries.

Red raspberries are the most familiar, while yellow raspberries are still rarely seen in stores because they are so fragile. However, both red and yellow raspberries are very cold tolerant, hardy as far north as Canada. Both the red and yellow varieties produce new canes from their roots.

Black raspberries have a slightly different aroma and depth of flavor compared to red raspberries, and many people agree that their flavor holds up better when cooked. Black raspberries and purple raspberries are not as cold tolerant as the reds and yellows. Black raspberries produce new growth when their long canes droop and make contact with the soil. The tips then form roots and grow into new plants. Purple raspberries are a cross between red and black varieties. Most purple varieties are similar in flavor to red raspberries; however some purple varieties produce canes, and others share the growth habit of the black raspberries.

The Raspberry Patch

Purchase bare root or potted raspberries that have been certified as disease-free. Potted plants are more expensive but have developed a stronger root system and will produce more new growth sooner than bare root plants. Although raspberries are not extremely picky about their soil, they do like good drainage, and of course mixing in compost as you prepare the bed is always a good idea.

Raspberry Growing Needs

- Well-drained, nitrogen-rich, composted soil
- Full sun
- Support structure or trellis
- Space for expansion

Raspberries grow aggressively, and they have thorns. Containing and managing their growth is a major task. As the plants mature they produce more canes and need more pruning. Raspberries are usually grown with the support of a trellis. While you may not need the trellis the first year, it is easier to erect it before planting your raspberries. You will want to erect two sturdy posts, one at each end of your raspberry row, and tautly suspend wires at about two and five feet above the ground. The posts must be anchored securely in order to support the considerable weight that the raspberry plants will place on the wires. Sink each post into the ground at least two feet and set with concrete. Additionally, you may want to anchor each post. Of course the

LEFT: These raspberries are properly trellised with three wires and wide rows.

length of your row depends on how big your raspberry patch will be. Red and yellow varieties should be planted two feet apart, black and purple varieties four feet apart. The space between rows needs to be adequate to allow you to prune and harvest, so plan on at least eight feet. Mulch your raspberry patch to avoid constant weeding among the thorny canes.

Planting Raspberries

If you are planting bare-root plants:

- Plant in spring or fall
- Soak roots for two hours in a bucket of water
- Spread the roots as you set each plant in its hole
- Backfill the hole with loose soil with crown just below ground level
- Gently press down soil
- Cut back the tops to two inches above the ground
- Pour one gallon of water around each plant
- Potted plants can be set out spring, summer, or fall, and need not be cut back; simply remove any damaged canes.

Pruning and Maintenance

Each year after the berry harvest, remove all of the canes that bore fruit, then prune away any spindly or sickly looking canes. Once your bed becomes established (about three years) remove new canes leaving no more than five for each foot of the row. In late fall cut back the canes to four feet tall. In the springtime remove any broken canes, tie canes to trellis, and apply fresh mulch to the patch.

Harvest your raspberries carefully. Make sure they are completely ripe—they will not ripen further once picked. Handle the berries gently. Place in shallow containers to avoid crushing them, and store in the refrigerator for no longer than a couple of days.

Blackberries

Blackberries have similar growth habits to raspberries; however, most are not as hardy and will overwinter only down to zone 6.

Blueberries

The most important thing to know about blueberries is that unless the pH of their soil is low (acidic), the plants will fail to thrive. We can personally attest to this as we have had a half dozen blueberry plants languishing in a corner of our

garden for four years. Our most abundant blueberry harvest was a mere handful of berries. So, before planting any variety of blueberry bush, do a quick soil test with a kit from the garden supply

Young blueberry plants at Poindexter Farm.

store. In most cases you will need to amend your soil by adding sphagnum peat moss, composted pine needles and oak leaves, and possibly powdered elemental sulfur. You want to bring the pH down to 4.5 to 5.0 before planting.

Blueberry Growing Needs

- Well-drained, acidic soil
- Full sun
- Space for expansion

There are four types of blueberries from which to choose, and your choice will depend on your hardiness zone and how tall you want your blueberry bushes to be when they reach maturity. Within each type there are many different cultivars to consider, each with unique attributes.

Low-bush blueberries are the hardiest of all blueberry types. These are the blueberries found in the wild and usually grow to be about two feet tall. Plant one to two feet apart.

High-bush blueberries have been bred to reach heights of eight feet or more and produce large berries. They are less hardy than the low-bush. Plant four to five feet apart.

Half-high blueberries are hybrids. They combine the coldhardiness of the low-bush, and the larger berries of the high-bush in a plant that grows three to four feet high. Plant two to three feet apart.

Rabbit-eyes are the blueberries to consider if you live in the south. Most blueberries need a good, cold winter, but rabbit-eyes prefer a winter that is simply chilly. They are heat and drought tolerant, but in very hot climates they do require irrigation. Their name comes from the pinkish hue of the fruit. Plant seven to eight feet apart.

Blueberry plants can be purchased potted, balled-and-burlapped, or bare-root. Bare-root plants should be planted as soon as the last hard frost has passed. Potted and balled-and-burlapped plants can be planted at any time during the growing season. Blueberries require an exchange of pollen in order to produce fruit. Encourage cross-pollination by planting more than one variety of blueberry. After planting, trim away any dead or broken wood, and remove any flower buds to prevent premature fruiting. Once high-bush and half-high varieties get established (after three to five years) prune not only broken and dead branches, but also overlapping and tangled branches. Wild low-bush blueberries should be pruned only every three to four years, at which time you should remove half to two-thirds of their branches.

Blueberries are ripe when they fall readily into your palm and burst in your mouth with sweetness.

Growing Mushrooms

Mushrooms are everywhere. Once you catch the mushroom bug, you'll be spotting them everywhere you look. After attending a book signing for a mushroom field guide in Boulder, Colorado, we were overwhelmed on our next mountain hike by the beautiful multi-colored forms that seemed to be poking up from the forest floor. Our eyes were opened to an incredible parallel universe that had been there all along, just out of our awareness.

Mushrooms are the fruiting bodies of fungi, which are in their own kingdom alongside plants and animals. The importance of fungi to soil health cannot be understated. They are the primary decay engines in our world. Fungi convert dead organic matter, such as wood, grass, and leaves, into nutrients that they use as food and distribute throughout the ecosystem. They do this through a "root system" called the mycelium, made up of "roots" called hyphae. This root system both enriches soil and creates microscopic tubes that allow air and water circulation to the roots of plants. Hyphae also form symbiotic associations with plant roots that allow plants to increase their nutrient and water uptake dramatically, as well as helping them to resist plant diseases and pests. Mycelium in the soil forms the glue that holds soil particles together and creates tilth. Healthy soil with many filamentous fungi can support healthy plants. They are another building block of healthy soil that discourages weed growth.

Perennials, which stay in one spot for many years, benefit from this relationship the most. Your garden will be more successful if you promote fungal growth in your perennial beds. The simplest way to do this is to mulch your beds. Fungal growth will begin naturally and instantly. Soon you can move aside your wood chips or dead leaves and see the tell-tale signs of fungal growth. You have probably seen these often white root-like hyphal structures under decaying leaves or logs. It's this mass of mycelium that eventually "fruits" into the mushrooms that you see and eat. The mushrooms fruit, then spread their spores (their seeds), and the life cycle continues.

Growing mushrooms is one of many complementary enterprises that hobby farmers can develop that creates a loop of beneficial activity on your farm. In this instance, the fungi consume dead plant matter, distribute nutrients and water

RIGHT: Mark Jones of Sharondale Farm demonstrates how to build a simple mulch bed for growing garden giant mushrooms.

throughout the soil to the roots of your plants, and expand your soil structure, all the while producing delicious and nutritious mushrooms for you to harvest and enjoy or even sell.

Cultivation

As a beginner, you'll want to obtain your starter material from a reputable mushroom grower (we provide some suggestions in the appendix). Mushroom spawn is available from a source listed in the appendix. Spawn is the mycelium that has been grown in sterile conditions on materials like wood or sawdust or sterile grain. Suppliers also have a lot of valuable information and suggestions about what mushrooms to grow where and their various methods. In very northern climates, it's best to start cultivation in spring. But most everywhere, you can start in fall or spring. Avoid the heat of the summer or your mushroom bed or logs will end up drying out before they've had time to become established.

The easiest method of cultivation is to create a mulch bed in your garden among your perennials with the garden giant mushroom. Lay down a layer of cardboard (avoid cardboard that is lined with color lamination, and remove the plastic tape and document pouches or labels). Then use the lasagna method and spread several inches of wet wood chips, some spawn next, more wood chips, spawn again, and so on until you've used all the spawn (two layers of spawn is sufficient). Then you can top it off with another mulch layer of cardboard and wood chips or just more wood chips, leaves, or straw to keep it moist. Soak it well with water and the mushroom bed is on its way. Keep watering occasionally until you notice the white mycelium growing through the chips or if the bed starts to dry. No need to do anything else except harvest the mushrooms when they come.

Mushroom logs are the most popular method of cultivating mushrooms for hobby farmers. They not only grow some of the tastiest mushrooms, like shiitake, but they look cool either racked up in rows, used as borders around perennial beds, or sunk into the ground

a third of their length like totem poles. They're an aesthetically pleasing addition to the garden, while they take years to break down into your soil, providing you with the long-term benefits of rich soil even after they've finished producing delicious mushrooms.

Oyster mushrooms are the easiest to grow on logs for the beginner as they don't require handling once they are inoculated and placed in contact with the ground in the shade. Shiitake need a little more attention, require you to soak the logs occasionally, and are less forgiving.

Cultivating Oyster Mushrooms or Shiitake on Logs

Logs—You need hardwood logs about three to four feet long and three to six inches in diameter. If they are too thin, then they might dry out too quickly and if they are too thick, they will be difficult to handle. The logs should be cut when the trees are dormant (or almost dormant) because more sugars are stored in the wood for the winter instead of in the leaves. The bark needs to be intact and not damaged or the log may dry out. Softer hardwoods, like poplar, ailanthus, or maple should work for oysters. Oak can be used for oysters, but it's best to save oak logs for your shiitake. If you buy a five-pound bag of spawn, you can inoculate twenty to twenty-five logs. If you don't have that many logs, you can use the leftover spawn to inoculate a mulch bed of straw or wood chips (shiitake won't grow as readily in a mulch bed).

Spawn—Buy it from a reputable source (suggestion in the appendix). For log inoculation you can buy it in sawdust form or in plugs that you can hammer into the holes you create. Loose sawdust seems to offer the best success rate, but plugs can be inserted with only a drill and hammer and don't require the more specialized sawdust tools.

All of these materials can be bought from various mushrooming suppliers you can find online, usually together in kits of various sizes. See box (page 104) for list of supplies.

Step-by-step directions for creating a mushroom log are opposite:

1. Drill half-inch-deep holes in a diamond pattern around the logs with holes about six inches apart along the length of the log.
2. Use the dowel tool to pick up the spawn and then pop it into each hole. Make sure the spawn completely fills the hole up to at least an eighth of an inch from the top. If using plugs, insert the plug into the hole and gently tap it in with a hammer.
3. Use melted wax to seal the holes with the wax applicator. The wax seals in the moisture.
4. Label the ends of the logs by nailing on the aluminum tags. Record the mushroom strain and date, as well as the tree species, on the tags.

Oyster logs can be laid directly on the ground and used as borders for garden beds, as long as they are in a shady spot. Shiitake are best stacked with the lean-to method, which keeps one side at least a foot off the ground. Or you can crib stack them like a log cabin several logs high.

Once the mycelium coalesces and is mature in the logs, mushrooms will form when the conditions for fruiting are right. Shiitake logs many times need to be

soaked before their fruiting time, commonly after they've already produced a flush of mushrooms and have been allowed to rest for at least six weeks. To soak them, submerge the logs in a big tub of water and weigh them down with something heavy for twenty-four hours. Soak the logs when the mycelium is mature. With oyster mushrooms, you don't have to soak the logs. A mushroom fruiting after the fall rains is a good indicator of when logs are ready to produce mushrooms for you. Whether you soak or not, it's a good idea to water your logs like you do your garden to keep them moist.

You should begin to see a small mycelial mass poking from the holes after a good rain—these are called primordia. They will quickly grow into mushrooms and

Mushroom logs supplies

· Hardwood logs in three- to six-inch diameters.
· Spawn in sawdust or plug form.
· Drill or a high-speed angle grinder.
· Mushroom drill bit or a half-inch drill bit with a diameter that matches your dowel or plugs.
· A dowel tool to insert the spawn into the drilled holes (or a hammer if you've bought plugs)
· A block of food-grade wax
· A camp stove to heat the wax. (Always keep a close eye on the wax to make sure it isn't too hot as it can catch on fire. A bit of white smoke is okay, but if it turns dark, lower the temperature.)
· A wax applicator
· Metal durable labels

you should harvest by slicing them off with a clean knife near the base before their caps have completely opened. Get them before the bugs do. There are too many variables to give an estimate on how many mushrooms you'll get per log (weather, moisture, inoculation rate, log size, etc.), but you can be confident that the mushrooms will do what they're designed by nature to do; and although your success rate may vary, you're still creating a new beneficial organic system in your garden.

You can usually get three to four "flushes" of mushrooms a year over three to four years before the log will be spent. Shiitake logs can be soaked from spring to fall once the log is colonized and can be soaked again every six to eight weeks during the season. After the log is spent, it eventually breaks down into mulch for your garden. Or you can line your woods path with them for serendipitous surprise mushroom growth. You can watch the natural succession of fungi breaking down the wood into mulch.

Once you've made some homemade pizza loaded with your own oyster mushrooms, or after you've had that taste of your own fresh shiitake mushrooms sautéed with garlic and butter alongside a fresh piece of fish from your pond, you'll be hooked. Soon you may want to branch out to include lion's mane mushrooms grown on oak logs or on mushroom bags in your basement, or almond portobello grown in compost beds. All the while, you'll be improving your soil and fostering another cycle of life on your farm to the benefit of everything on it, including yourself.

BELOW: Oyster mushrooms ready for the frying pan.

CHAPTER 6

The Flower Garden

Growing Flowers for a Cutting Garden

It's impossible to be surrounded by land and not want to do something with it. Our house was surrounded by two and a half acres of grassy and weedy lawn. Built in 1935, the only foundation plantings were trees. None of the previous owners had done any type of landscaping—there were no bulbs, flowering shrubs, or perennials.

I wouldn't say that living surrounded by viable farming land inspired me to quit my teaching job, but wanting to do something that utilized this land in a positive way did occupy my thoughts after I had decided to take a break from teaching. Since we first moved onto our bit of land, we had wanted to participate as vendors at the local farmer's market. We knew we could grow things, but with both of us working full time we could only manage a kitchen vegetable garden and the various beds of flowers we planted around the house.

As the idea of using our land spun around slowly in the back of my mind, touching on various plans and projects (spinning our llamas' fleece into yarn, growing herbs for essential oils to make skin care products, keeping bees and selling honey . . .), I was intrigued by my sister's venture into growing and selling flowers. She and a couple of friends went in together on the cost of bulbs and the labor of planting and harvesting. Unfortunately the endeavor fell apart when one of the partners stopped helping with the labor, which then spiraled into a collapse of the venture. And so I began to think about growing flowers to sell at the market as a way to make use of my land.

I've always loved flowers, and I'd had a fair amount of luck with the flowers I'd planted around our house. I began to research this flower growing business. After checking out The Flower Farmer from the local library and reading it through, I was hooked. The next step was approaching Michael with the idea and getting him to invest both the money and the time to make the project happen. This was much easier than I anticipated. I began making a list of flowers that would do well in our climate and were considered good cut flowers. I ordered seeds and started them inside on heat mats with lights. Michael and I attended a cut flower workshop sponsored by the agricultural cooperative extension agency and this really helped us gain a perspective on what we were getting ourselves into. We acquired much useful information, especially in the form of resources to access for various supplies (drip irrigation, row covers, support netting) and met with other local flower growers, one of whom had been our local market's biggest cut flower vendor and is still active in the market community. —Audrey

Having a cutting garden is a much different situation from having a landscaped garden. When we tell people we grow cut flowers to sell at the farmer's markets, most of them reply, "Oh, you must have a beautiful garden!" Well, yes and no. It looks great before we go out to pick, with literally hundreds of blooms topping lush green foliage. After picking, the garden is simply rows of bare greenery and undeveloped flower buds. Imagine a row of 500 gorgeous, colorful, stately tulips. After we've been out to pick, only the pale green leaves remain. In contrast, a landscaped garden is there to provide visual pleasure and a specific type of structured space to enhance your lifestyle year round. A cutting garden is all about how to grow the most beautiful healthy flowers the most efficiently and, in our case, in the most environmentally friendly way possible. In short, we don't just grow flowers, we farm them.

Planning and Constructing the Cutting Garden

If you are planning a personal cutting garden, you can adapt most of what is discussed in this section to fit a smaller scale and your own personal needs. As always, look at what you have available to you and try to adapt it to work for your purposes. A southern exposure, allowing full sunlight for six to eight hours a day,

RIGHT: Orderly rows being constructed in spring, before the weeds invade, using landscape fabric and mulch for the pathways.

is necessary for growing most flowers. We were fortunate to have three quarters of an acre adjacent to our vegetable garden and already fenced off from our pastureland. With minimal additional fencing—we used chicken wire and cedar posts cut from our woods—we excluded this space from the dogs and the chickens, and then built a sturdy gate wide enough through which to fit the lawn tractor and a large wheelbarrow. The other gate was wide enough to fit the tractor and truck. Just as with your vegetable garden you will need access to a plentiful water source. For the long straight rows we planned for the flowers, drip irrigation is the most efficient watering option, both in terms of water use and time management. Drip irrigation is also convenient, easily adaptable to your planned space, and fairly inexpensive.

We planned on five long rows of flowers to fit the space and still allow room for the tractor to maneuver—two 3x60-foot rows, and three 3x80-foot rows. Our first year we learned the hard way that the paths between the rows need to be at least four feet wide, five if you have the space. We could barely squeeze through the zinnias and cosmos on either side of one path, and there were certain spots that became inaccessible as the season wore on. It is difficult to imagine how big those tiny little seedlings are going to be in only two months' time! The following year we were able to expand the paths for all but the first of our rows.

In order to prepare the land we invested in a plow for our tractor and a tiller. Neither of these tools is recommended for soil quality and should be used very sparingly. But if you're starting with a field of heavy grass and weeds, there's really no other option to initially prepare the ground for planting unless you want to spend several years working it by hand. Of course, borrowing these tools for the few times you need them is also a good option.

Nothing productive had been grown in this space before and the ground was mostly compacted clay and rocks. That first year we brought in two dump truck loads of soil mixed with compost and added it to the beds as we tilled. As we picked rocks from the soil we also came across old glass bottles, pieces of cinder block, bullet casings, fragments of dishware, old nails, and twisted pieces of metal. Even as we work these beds three years later we still occasionally find these things in the soil. In addition to the human junk, our old nemesis the wire grass had a firm hold on much of this space. While it is still a nuisance, we have managed to grow and cultivate flowers in this area. They are mostly annuals that we till up each year, rather than perennials that the wire grass would eventually envelope and suffocate.

How to Choose the Right Blooms for Your Farm

Perennials

Perennials, planted in the correct hardiness zones (see page 86), will regrow each year from their roots. It is important to carefully consider their placement in the garden since they will require a permanent home and can become very large in size. A few examples of commonly grown perennials used for cut flowers are daisies, echinacea, peonies, and yarrow. In our cutting garden the perennials occupy the first row, closest to the vegetable garden, where it is now impossible to maneuver the tractor. As our business grows and our garden aspirations expand, we are beginning to extend perennials into the second row that we have been using for bulb flowers. Perennials are like the backbone of the cutting garden. By planting an interesting selection of perennials that bloom over a long period of time, you provide yourself with a base crop of flowers that does not require you to seed and cultivate the plants each year. Perennials are an investment, both in time and money. Unless you grow your perennials from seed, and this requires at least one to three years for the plants to become mature enough to produce large and abundant blooms, you will need to purchase plants from a reputable source.

Be selective and plant what you really want; remember, these plants will be around for a long time, and moving them is not an easy task. We grew eight echinacea (purple coneflower) plants from seed our first year of flower growing. At the end of the first season we got a few small blooms. We were looking forward to the second year's crop of these bright, stately wildflowers. Apparently, so were the Japanese beetles. The plants bloomed gloriously, and the beetles decimated all but one or two blooms before I could harvest them. In an effort to get some return on our time and effort, we picked the stems and removed the tattered, dirty leaves from the center seed cone. We used these bright orange and rusty brown stems in

BELOW: A row of lisianthus grows at Green & Gold Farm.

bouquets with great success. However, the following spring, we found that the echinacea, being a wildflower, had self-seeded and taken over a third of the row in which it was planted. We tried to dig it all up and move the plants to various spots in our regular yard, but even as we sit here, echinacea plants are growing in our spring bulb row.

Woody ornamentals, perennial trees and shrubs, are another extension of the cutting garden. Although most "woodies" are traditionally landscape plants, they can serve double duty by providing interesting cuts to bring indoors. Examples of commonly found woody ornamentals are buddleia (butterfly bush), forsythia, hydrangea, and lilac.

Bulbs, Corms, and Tubers

Bulb, corm, and tuber flowers are an upfront expense and can be quite costly. Bulbs also require extra digging, which means extra time and effort, to plant them at the required depth. The advantage of planting bulbs is that you then have flowers early in the season. In April, people are eager for a bright burst of color after the bleak coldness of winter. We got a late start that first year. We didn't really begin to plan out our flowers until December, by which time it was far too late to plant bulbs. We didn't have enough flowers to attend a market until the very end of June! Bulbs do take up quite a bit of space, but in many cases (i.e., tulips) growers till them up after harvesting them, thus treating them as an annual crop. This makes sense if you need the space for other crops that season, but it cuts your profit margin drastically. In general, most bulbs whose flowers are cut will not return with the strength and vigor of bloom they produced in their first year.

Most tuberous, or rhizome, flowers are more expensive and are meant to return year after year. Examples of these are dahlias and tuberose. They do not have to be planted as deeply as most bulbs, and many have the advantage of producing more than one bloom per tuber. An important consideration when selecting bulbs, corms, and tubers for your garden is whether or not they are hardy enough for your climate and can be left in the ground over winter. Depending on your hardiness zone, some bulbs and tubers may need to be dug up and stored in peat moss or sawdust over the winter months. If you must store bulbs and tubers over the winter make sure you have a dry, dark space that is protected from rodents. Some bulbs, corms, and tubers will naturalize—return each year and even multiply. Check the hardiness zone for each variety. If you do leave your bulbs, corms, and tubers in the ground over the winter, be aware that they make tasty meals for voles and other critters.

Annuals

The greatest volume of flowers in a cutting garden are produced from annuals, those flowers that live only one season and must be replanted each year. Prime examples of these are larkspur, sunflowers, and zinnias. How much space you devote to each variety of flower depends mostly on the size of the plant at full growth, and how many of them you need and want for your markets. As always, much will be learned through experience. Our first year of growing cut flowers we woefully underestimated our customers' desire for giant yellow sunflowers. Luckily we had the space in our garden and were able to plant 500 single-stemmed sunflowers the next year and we're up to several thousand now. Another mistake we made was not planting any type of ornamental grass or greenery for use in bouquet making. Using inexpensive grasses and greenery helps boost the value and adds texture and depth to your bouquets at the market. Plus, the more room in a vase taken up by greenery, the less flowers you have to use to fill it out.

Your annual flower garden begins with the arrival in the fall of that season's seed catalogs. Most seed companies now have separate sections devoted solely to varieties of flowers that are best suited for cut flower use. Ask around and find out where other growers get their seeds. Look in gardening magazines and online. Get on the mailing list for as many catalogs as you can so that you will be aware of all the newest varieties of flowers. Then make your wish list. If you are growing in the field you will need to pay particular attention to the hardiness zone of the flowers that catch your eye. One of my favorite flowers is the delphinium and I confess to feeling thwarted and constrained by the fact that I don't live in a climate conducive to producing delphinium. I've attempted to grow it anyway, and was finally

rewarded this year by a crop of about ten magnificent spires of delicate flowers. It makes sense to order the bulk of your seeds from one company. This way it is easier to keep track of your order and you'll have only the one shipping fee.

The size and variety of flowers in your cutting garden are entirely up to you. We try at least ten new varieties of flower each year, most of which don't work out that well and we don't plant again, but those that do are added to our list of regulars. If you are growing to sell at market, remember that customers are attracted to the new and unusual. Having something that no one else offers will bring customers to your market stand. Also consider how you will be marketing your flowers. If you do any kind of bouquet or arrangement work, having a plentiful supply of filler material is essential. Ornamental grasses, distinctive herbs, and interesting shrubs provide a low cost way to add body and interest to a selection of flowers. You can add value to bouquets by using several specialty flowers like lilies and then filling them out with less expensive flowers and filler. Decide on several foundation plants, those that you will grow every year, and then try out other varieties around these.

Elements of a Cutting Garden

Important elements of a successful cutting garden include row planting with wide pathways; a drip irrigation system; woven landscape fabric; and support netting.

Row Planting and Wide Pathways

Because irrigation is so important, rows are the best method for farming flowers. And it's just as important to maintain weed-free, wide pathways. Flowers can grow into very large plants that spill out of the sides of your rows and make it difficult to maneuver and cut the flowers. We maintain four- to five-foot rows where possible. We cover the ground with cardboard and a thick layer of wood chips in the pathways to keep weeds at bay.

Drip Irrigation

By far the most important element is the drip irrigation system, without which we could not feasibly provide water for the eight long rows of flowers in our half-acre cutting garden. Not only does the drip irrigation allow us to turn on the water and proceed with other tasks, it delivers that water right at ground level, allowing it to soak slowly into the soil, going where the plants need it most, right to their roots. This system prevents waste of water from evaporation and over watering, and also prevents diseases encouraged by the excessive moisture of overhead watering, such as powdery mildew. It's best to buy a kit from an irrigation supplier. Some recommendations are offered in the appendix.

Drip irrigation is relatively easy to install and disassemble.

LEFT: It doesn't look pretty, but drip irrigation saves you a ton of time. And when the plants grow in later in the season, you barely know it's there.

Installing Drip Irrigation

1. A pressure regulator and filter are needed for each drip system you have. We have two.

2. Main line tubing runs from the filter to the beds. Use right angle and t-connectors to get the main line to the beds.

3. Use a special drip irrigation tool to bore out holes in the main line for the drip tape connection.

4. Insert the drip tape valve into the main line and tighten.

5. Attach drip tape to valve and tighten.

6. Close the end of the drip tape by cutting off a piece about an inch long. Fold the end up twice and then insert the cut piece onto it.

RIGHT: Burning holes in landscape
fabric with a propane torch.

Landscape Fabric

Not all growers use landscape fabric
on their rows. Landscape fabric is a heavy
black woven polypropylene fabric that allows
water to percolate through to the plants. It
is available in several thicknesses, lengths,
and widths. We find it essential to weed control because we self-seed and set out
our own tiny transplants. The landscape fabric allows these baby plants to grow
without being overtaken by the hardy and abundant weed population. Through
experience we've learned that weeding around baby flower plants leads to as many
flowers being destroyed as weeds. We do not have landscape fabric down in our
perennial and bulb/tuber beds. We also do not use landscape fabric for our giant
sunflowers. The sunflower is one plant that grows quickly and strongly enough to
outpace the weeds. We do end up with a lot of weeding to do once the sunflowers
have established themselves.

An important distinction to make is that between landscape paper and landscape fabric. Our first two years we used landscape paper, and it worked fine. Unfortunately, it only lasted one season. The paper disintegrated and tore, making it
impossible to pull it up and save it for another season's use. While not our biggest
expense, the landscape paper is not cheap. Also, we had spent several hours marking
out the flower hole placement and then burning a small hole for each plant with a
small propane torch. Yes, thousands of holes. From another local farmer and friend
of ours, we learned that a local landscape company sold landscape fabric—the actual
synthetic woven material—and that is what we put down this year. We plan to pull
it up and store it over the winter and reuse it again next year. After cultivating and
composting our beds, we lay out our drip tape and then spread the landscape fabric
on top, fastening the sides with landscape fabric pins (sometimes called staples).
Not only does the fabric allow our new flowers a chance to grow, it provides weed
protection throughout the summer. On either side of our rows we construct paths
by laying down cardboard and covering this with wood chips, mulch, or straw. With
this method we have managed to squelch the majority of our weed problems, which
allows us to focus our time and attention on our plants.

Support Netting

Strong wind and heavy rain can quickly destroy a crop of tall flowers by
bending and breaking their stems. We use support netting in our long rows of
flowers to help prevent this type of damage. Support netting also helps keep the
plants growing in an upright habit when their blooms become heavy. Support
netting is a plastic mesh with approximately four-inch square openings. It is

often sold as a trellising material and comes in different lengths and widths. It is easily cut to fit any size bed, and is sturdy enough to store and reuse season after season.

Support netting is available through gardening catalogs and garden supply stores. Prepare your bed—apply compost and landscape paper or mulch, lay drip hose, and plant—then erect your support netting. You will need posts or stakes that are tall enough to match the flowers' growth, and sturdy enough to hold the netting in place during severe weather—rebar works well. Place posts at the four corners of the bed, and then every six feet or so along the sides of the bed. After cutting your netting to the proper width and length, slip each edge corner over its post on one end of the bed and unroll the netting along the row, slipping a netting square over each post as you come to it. Lower the entire row of netting to the level of the plants by sliding it down the posts. As your flowers grow they will grow up through the mesh and you can raise the mesh to support the stems. Some tall flowers need two levels of support netting. You'll want to put the netting in before the plants grow too tall. Not all flowers need support netting. Often the seed catalog or packet will supply information about netting or staking. Our single stem sunflowers never need staking as their stems are the size of a small tree.

BELOW: Support netting keeps your flowers straight and protects them from heavy wind.

Cutting and Storing Flowers

Taking care to practice proper cutting technique and cleanliness will give your flowers a longer vase life. The vase life of our flowers (sometimes up to two weeks) helps distinguish us from other growers at the market that sell other products and don't have the time to properly cut, store, and care for the flowers they sell.

✓ Flowers don't like to be cut in the heat of the day. Either cut in the morning or later in the evening. Some flowers prefer either morning or evening and there are many books that will provide you with proper cutting times for individual flowers. The morning is the time when flowers are the most hydrated. And the evening is when they perk back up after a long day of absorbing sun.

✓ You should always use very sharp specialty flower shears that are very clean. Dull shears will crush the stems and damage them. Cutting them cleanly also allows them to absorb water more readily. Any bacteria on the blades of your shears will transfer to the flower and decrease its vase life.

✓ Some flowers, like zinnias and hydrangeas, require that you remove all the foliage as the leaves will suck water away from the bloom. Again, you'll need to do some research on each type of flower you grow to find out the proper cutting technique. If you need to take time to remove the foliage on a stem, then make sure to re-cut it at an angle before putting it in your bucket.

✓ Use very clean buckets when harvesting and storing. Always wash out your buckets with soap and water after use and store then in a clean place. Once you've filled your bucket, either store in a cooler (we bought some old drink coolers from a convenience store that went out of business) or in a cool, dark place like a basement. Don't put any fruits or vegetables in the same cooler. In a pinch, a cool, dark room of your house will work, as long as you're planning to sell the flowers within 24 hours.

✓ Many people use flower preservatives in the water. We don't. We sell our flowers so quickly that we rarely have them sitting around for long. And we've never had trouble making our flowers last at least ten days without using preservative. But experiment with it yourself and make your own choice.

✓ When the flowers are in a vase, make sure to keep them watered and change the water every couple of days. This is a good thing to tell your customers to do too. And don't forget to constantly clean your buckets, shears, and vases after every use.

Farmer Profile

Green & Gold, LLC., Flowers by Eileen Stephens, Earlysville, Virginia

After her grant money dried up at her university job as a chemist, Eileen Stephens asked herself, "What would I do if I could do anything in the world that I wanted?" Her answer was to spend most of her waking moments for the last twenty years in her extensive cut-flower garden.

She and her husband had lived in their house on several acres outside of Charlottesville, Virginia, for twenty years when she made her career change. Over the years her husband had established and nurtured a thriving vegetable patch, but there was plenty of space for Eileen to create her cut-flower garden. Eileen turned the plot of about a half acre solely by hand, proving the point that you need not make a huge capital investment to begin pursuing your dream.

Initially, the rows were laid out to match up with the rows from the vegetable garden, but soon she realized they needed to be rotated 90 degrees, running north to south so the sun would hit the plants evenly throughout the day and none would be shaded by taller plants. Working on her own, Eileen has produced a thriving perennial garden, watered with drip irrigation, that includes about 500 peony plants. She has taken over most of her husband's vegetable garden and it now contains rows of meticulously cultivated lisianthus. As she walked us through her garden, giving us the "historical tour," she pointed out the many different types of hydrangea she's acquired over the years. Gradually she expanded from one garden plot to four, taking advantage of partially shaded areas on her property to grow her gorgeous helleborus. The last garden plot she showed us is an annual garden that includes, along with zinnias and sunflowers, rows of tuberose and dahlias that she digs up each fall to replant in spring. Several years ago she added a greenhouse in which she starts her seedlings and grows thousands of lilies in crates over the summer months.

She cuts her flowers throughout the week and stores them in a three-door cooler before she takes them to the local farmer's market on Saturday mornings, where she's been a fixture for twenty years. Peonies are her best sellers at market. She also sells her flowers to two local florists. It took Eileen five years to become profitable and she still doesn't make a full living from it. She began as a hobby farmer and has built up her business over a couple of decades to be profitable. But she says, "It's worth doing, if you enjoy it." She has been asked if flower farming is a full-time job and her reply is, "No, it's three full time jobs!" She has employed help for about the last three years, but she reports that the increased volume she gets from using helpers only makes up just enough to employ them. Still, she loves to pass on her knowledge to other aspiring growers.

In her early sixties now, Eileen is beginning to slow down the pace of the business. Going forward, she'll only sell her early spring flowers at market and will avoid the intense heat of the triple-digit days that come later in the summer. Her hands show the experience and wear of a true gardener, and she credits her good health and spirits to the years working her soil.

Eileen has two pieces of advice for those getting into a cut-flower hobby farming venture. One is to join the Association of Specialty Cut Flower Growers (www.ascfg.org). They provide education and a ton of resources for growers. Their bulletin board is a terrific place to troubleshoot. And you'll find lasting friends and partners with whom you can network and share info and resources.

Her other piece of advice is never to plant peonies, a perennial spring flowering plant, in a spot where you've grown them before. It won't work, even if you wait seven years and plant other things in that spot in the meantime.

CHAPTER 7

Extending Your Growing Season

In order to grow flowers and vegetables for the entire selling season from April to September, you'll need to employ several strategies. Useful tools that can help extend the growing season include cold frames, row covers, and hoop houses or greenhouses.

Growing for the Selling Season

If you're growing flowers for the market, you'll want to plan accordingly so that you have flowers from the first market in April to the last market in September, or whenever your local farmer's market runs. This is hard to do and we still struggle to have a full selection for the entire season. But we come closer to reaching that goal every year. All of the techniques for extending the season discussed here are basically tools for you to get an early start on the season. Flowers are in big demand after a long, cold winter. But don't forget to plan for the entire summer. You'll need to succession plant your flowers carefully (go by the days to germination and time of bloom information on your seed packets and in the plant catalogs) so that you know what will be blooming and when. Don't get too bogged down your first year in this process. Just grow what you can, when you can. But you'll learn from experience what is in demand throughout the season and you can refine your planning and planting accordingly.

LEFT: Green & Gold.

Getting an early start on your garden has many benefits. A cold frame or row covers allows you the opportunity to enjoy leafy greens as early as eight to ten weeks before the last hard frost of the season. An added benefit of this early growing time is the absence of most garden pests. Many crops can tolerate a light frost, and there are those crops, such as cabbage, broccoli, Brussels sprouts, and those yummy leafy greens, that benefit from an extended spell of colder weather. Cold frames and row covers give these crops a chance to thrive in the conditions they like best. Cold frames and row covers also allow you to extend your season into late fall and winter so you can continue to enjoy homegrown vegetables past the first hard frost of autumn. To find out when to expect your first and last hard frosts check the national climatic data center website, which is listed in the appendix.

Cold Frames

A cold frame is simply a box frame with an angled, removable, or hinged top that lets in light. Traditionally made from wood, they can be as big or small as you want to make them. The top can be glass or plastic—old window frames can serve this purpose; just make sure they don't contain any lead-based paint or putty. The top is set in at an angle to let in as much sunlight as possible. The purpose of a cold frame is to provide a friendly growing environment and protect your plants from freezing. Place your cold frame over a plot of prepared ground so that the top is facing south, allowing for as many hours of sunlight as possible. Plant your seeds and check on them often. It is important to vent your cold frame (prop open the top with a stick) to allow air to circulate around your plants. On warmer days, open the lid wider so that your plants do not get too hot; the small, enclosed space of a cold frame can easily overheat.

If you are starting off with a very small garden, you can even use plastic water jugs with the bottoms cut out to make mini hotbeds for individual plants. Simply place the bottomless jug over your small plant to protect it from frost. Take the cap off to provide ventilation on warm days, and replace it in the evening.

Row Covers

In the cutting garden, row covers make it possible for us to get our snapdragons and sweet peas planted outside during the cool weather that these flowers crave. Row covers are lightweight lengths of polypropylene fabric manufactured in different grades of cold protection. The row covers capture warmth, encouraging earlier yields and healthier plant growth. For many crops, row covers may be loosely directly spread over the crop and secured at the edges with row cover tacks or staples. For those crops that need the space to grow up, use row cover supports, or hoops to create a "low-tunnel." Because the row covers are sunlight, air, and rain permeable, the plants are able to grow without having to constantly remove the cover. Along with an extended growing season, row covers offer the benefit of pesticide free, non-toxic insect control and protection from damaging winds.

Hoop Houses and Greenhouses

For the more serious grower who is ready to make a bigger investment in extending his or her growing season, there are hoop houses (also called high-tunnels), and greenhouses. A hoop house is a plastic-covered structure that is high enough for a person to stand up in and walk through. It relies on solar heating and manually adjusted ventilation. Hoop houses are also considered

BELOW: Our hoop house under construction.

Some things to consider before investing in a hoop house or greenhouse:

- What are the permit requirements and construction specifications for your area?
- What will you be growing in the structure and for what purpose?
- Do you have the space for a permanent or semi-permanent structure?
- Do you have the means to supply water (and heating or cooling systems) to your structure?
- How much will you actually use the structure?

semi-permanent structures and can be built on skids to allow them to be moved seasonally. A greenhouse is permanent or semi-permanent and has artificial heat and venting systems. Both of these structures can be built as large or as small as the grower desires.

Once you've answered these questions you will be better prepared to make a choice about what kind of hoop house or greenhouse you need, and what is feasible for your situation.

Small hobby greenhouses can be purchased as kits and are fairly inexpensive. This is a good option if you lack the space to start your own seeds indoors. Audrey's sister, Elizabeth, starts hundreds of herbs, greens, onions, broccoli, cabbage, squash, eggplants, and peppers each year in her 8.5x8.5-foot hobby greenhouse. Having this structure allows her to grow all of her own healthy, organic seedlings, providing food for her family and an abundant selection to offer at her local farmer's market on the West Coast. Small structures like this that can be easily disassembled do not usually require a permit.

If your intention is to grow crops through the colder months, you'll need a structure large enough to accommodate several beds of vegetables or flowers. Depending on your climate and your crops, an unheated hoop house may serve your purpose. As our society becomes more concerned with how its food is grown, and where that food is coming from, more information and more products have become available to assist those who desire to become part of a community of sustainable agriculture. There are excellent books and worthwhile websites that cover building your own hoop house or greenhouse. In addition, several mail order companies offer an enormous array of kits and products for constructing and expanding hoop houses and greenhouses.

It is this *process* of gardening, of learning about your land and your plants, that is so captivating. And when you can really begin to understand the relationships that exit between all the living things in the garden, you'll truly feel like you've become a part of it yourself.

The vegetables and other crops in your garden are just one part of its growth—the most visible and so the easiest to judge. But you'll soon begin to notice more earthworms, or wonder why there aren't more earthworms. You'll notice if mushrooms are popping up around your perennials. Does the soil in one bed of your garden drain better than a bed right next to it? Why don't tomatoes ever grow well in that one spot near the rose bushes? These are some of the questions and revelations that inhabit the dreams of a happy hobby farmer.

BELOW: Using row covers and straw as mulch helps encourage blooming of the earliest flowers.

The Care of Living Creatures

CHAPTER 8

General Animal Care Basics

Our Philosophy of Raising Animals

We'd already been practicing vegetarians when we made the decision to begin farming. Coming from the beautiful mountains of Colorado where we met, we have always been concerned with the impact that eating meat has on our natural world. It takes approximately twenty pounds of grain (which is how most of the world's animals are fed for meat production) to produce one pound of beef. That is not a sustainable ratio for the environment if we eat meat every day. In addition, most people are completely divorced from the source of the meat they consume. Even the grass-fed farmer at your local farmer's market has his animals slaughtered and butchered by someone else (due to USDA regulations). And forget about trying to rationalize the commercial meat industry; as soon as an industrial production model is attached to any living sentient being, it's inevitable that injustice will be done. Our philosophy is that if you can't kill and dress your own meat (not every single bit you eat, of course), then you aren't honoring that animal and you aren't honoring yourself. But after moving to a farm and raising animals, we quickly realized that achieving a level of ethical purity was close to impossible. So while we still don't consume meat, we understand that many people want to raise and consume meat in the healthiest and most ecologically sound way possible. Hobby farming allows for you to come the closest to that ideal.

We do eat fish. We've fished ponds for catfish, green rivers in Texas for bass, marshes in the Keys for red fish, mountain streams of Colorado for trout, and the Chesapeake Bay for rock fish. We've even ventured into the Pacific on a salmon fishing boat off the coast of Oregon that fishes with lines, not nets. We've caught, cleaned, cooked, and eaten our fish (but certainly not a majority of it) and feel a connection to it. Fish also live in the wild (we do not eat commercially farmed fish) where they aren't crowded together and fed antibiotics to keep their tight living situations from spreading disease. We follow the sustainable fisheries guidelines and stay away from species that are overfished. And fish is healthy.

Milk and cheese are another issue. You can't get milk without a cow, sheep, or goat that has been pregnant. And cows don't only give birth to female cows. They also produce bulls. And those bulls don't produce milk, so they are sold as meat. You can't get cheese without milk. So if you're eating milk and cheese, then you are also contributing to the meat culture. We get that. But we're also not completely dogmatic in our approach to food and meat, and we do eat milk and cheese sparingly.

There are the same ethical issues with eggs. You can have a small flock of hens for yourself and get eggs for many years and that's both good for you and the environment. But hens come from somewhere and that's usually a factory breeding operation. And the ratio of male to female chickens that hatch is usually 50/50, so if you have hens, then you contributed to the creation of an equal amount of cockerels that were then used for meat production. As with most green endeavors, there's really no way to

reach a state of environmental purity unless you foreswear keeping any animals at all on your farm. But that's no fun.

We're not so strict in our diets or ethics that we would turn away our grandma's homemade meatballs once a year or our friend's grass-fed beef once in a while. And we do understand that you can't go from a meat culture to a vegetarian culture, with the masses of people that need to be fed, overnight. We welcome our neighbors that raise their own meat for their own consumption and even those that sell at the farmer's market. Hobby farmers are an integral part of moving our society away from industrial meat. We will even give you the basics of keeping cows for grass-fed beef in this book.

There are many ethical questions everyone should explore when deciding to raise animals. Whether you make the decision to eat animals that you raise or not, you'll still want to make sure that you aren't creating an imbalance in your farm system. We practice an adoption and no-breeding policy. After all, does the world really need any more farm or domesticated animals? All our four dogs and five cats have been adopted locally from rescue organizations or they've just showed up as strays (live in the country long enough and animals find their way to you); they are all spayed or neutered. We have chickens, but we do not allow them to hatch eggs and breed. Our latest batch of chickens has come from another farmer friend who'd given up on farming and was selling his entire chicken outfit. We eat their eggs to prevent them from multiplying. And our donkeys and llamas are males. Males are always the least desirable animals for breeders (besides prized breeding stock) and we give the unwanted males a good home. We've stayed away from livestock markets because of the poor treatment and health of the animals. And we certainly support any organization that rescues animals from livestock markets.

This philosophy of not raising farm animals for meat also simplifies our life as hobby farmers. After all, time is tight when you're trying to balance an off-farm job with on-farm duties, and one of the easiest mistakes to make when getting into farming is to obtain too many animals. Animals take time and money. They begin to breed and before you know it, you're overwhelmed. If you're a hobby farmer and have a job outside of the farm, then low-maintenance animals are the way to go. The idea of homemade goat's milk cheese is divine, but just consider the time it will take to milk goats every day (you can't miss even one) and sometimes twice a day. That's why we chose donkeys, llamas, and chickens. They are all very self-sufficient and require little in the way of feed, if you have healthy pasture, and daily care.

But everyone feels differently and this is just our approach and our advice on raising animals and can be applied to raising them for meat as well. We'd recommend that you think long and hard before raising any animals at all. Resist the idea that you can only call yourself a farmer if you have farm animals. Farm animals live much closer to the natural world than humans, dogs, or cats. Many times they die in painful and tragic ways. And sometimes that death is not quick and it will be up to you to help the process along. You can't fully prepare for this, but certainly it should be a consideration.

Why have animals at all on a farm if you're not going to eat them? There are several reasons:

- Companionship—We enjoy being around animals, more so much of the time than humans. We consider our animals some of our closest friends. There are so many abandoned animals that need a home that it would be heartless to have all this room and not share it.
- Eggs—Eating eggs from your own chickens that you don't allow to breed is the most humane way to consume animal protein.
- Natural fertilizer—Our donkeys' and llamas' sole purpose outside of enjoying their lives is to convert our fields of grass to manure, which we in turn use as our primary source of soil enrichment for our vegetable and flower gardens. Our chickens serve this purpose too.
- Protection—Our dogs are a much better deterrent to intruders, both human and animal, than a gun. Long before we'd have time to get a gun out and confront someone or something, the dogs would have alerted us to danger and most likely have deterred any intruder from coming near our property. Our donkeys and llamas guard the fields from approaching predators.
- Pest control—If you live in the country, you will have rodents, including rats. Cats and dogs are the best natural form of pest control. And chickens eat loads of harmful insects, like ticks and Japanese beetles.
- Work—We don't work our animals, but we do know that our donkeys and llamas would be very happy to be included if we asked them to help pull logs out of the woods or carry a load on a packing trip.
- Sustainable by-products—In our climate, llamas need to be sheared once a year so they can endure the summer heat. Their fleeces are valuable and renewable each year. Similarly, sheep and certain goats provide fleeces that are also a sustainable way to use your animals without killing them. And if you truly care to milk an animal every day, then a single cow pregnancy can turn into a couple of year's worth of milk. Our beehive provides both honey and good pollination for our flowers and vegetables.

Animal Care Basics

There are loads of really terrific books out there that offer detailed information about raising and keeping farm animals. Our book will give you some general considerations and our own experience, but there are many people out there with much more experience than us. The other books in the *Joy of . . .* series are terrific companions to this one and there are others that are worthy, which are listed in the recommended reading section. We suggest you buy all the books you can find on the animals you will be raising because they all have something to offer. You can never get too many perspectives on raising animals.

Where to Find Animals

We recommend buying animals locally. If you're buying them as pets or as fertilizer producers, then there's really no reason to look for the pick of the litter. It's much easier and cheaper to find the gelded males that for-profit farmers don't care to spend money feeding. Ask around at your farmer's market. Ask the folks selling beef, chicken, eggs, llama wool, or sheep and goat's milk cheese if they have any animals they'd like to sell or know of someone who does. Don't be impatient;

you may have to wait until the next round of birthing. Just start putting the word out and the animals will come. We found many of our animals through the local rural classified newspapers that are available at most country stores.

Buying tips for first-time animal hobby farmers:

- Buy locally and befriend the sellers so you might be able to call them for answers to questions you may have later.
- Read up on the specific animals you're buying and the traits to watch out for, like founder rings on the hooves of horses. But unless you plan to show or breed the animals, there's no need to be too picky outside of general health.
- Always buy two to begin with as farm animals do not do well living alone.
- Begin by buying gelded males. They are cheaper and you can learn to care for them and give yourself time to decide if you want to breed animals in the future. Buying a breeding pair instantly commits you to more than your original purchase.
- Visit the farm you're buying from at least once without actually buying. Don't buy on the spot. Take a look around, ask lots of questions, then go sleep on it for a few days.
- Stay away from livestock markets. They can smell a greenhorn coming from miles away and you'll never leave there with what you truly want or need.
- While we do advocate adopting from rescue organizations, don't do it initially. Rescue farm animals can have very serious physical and mental problems that may be beyond the expertise (or financial means) of a first-time farm animal owner.
- Start with the lowest-maintenance animals you can find, like llamas, donkeys, longhorn cattle, or chickens. It's much more fun and rewarding to grow into a menagerie at your own pace than to have them take over your life.
- Remember that even chickens can live twenty years or more. Horses, llamas, and other large grazing animals can live well into their thirties. Keeping animals you aren't going to eat is a big commitment.

What Do All Animals Need?

Shelter—Most of the animals we recommend in this book need just a three-sided shelter with a roof to be happy. They'll only use it to get dry or stay out of the harshest weather. If you live in a very cold, wet, or hot climate or you're raising show animals, then you will need a barn to keep them during the coldest and hottest days of the year. (Chickens need a fully enclosed shelter to keep them safe from predators; more on chicken coops and tractors to follow.)

Food and water—The animals we recommend for the hobby farmer need very little in the way of extra food, as long as you give them enough pasture to graze. Animals are not healthy when they are overweight and you should not feed your farm animals in the same way you feed your dogs and cats. They should be left to forage mostly for themselves, except when that forage is not available. We only buy hay for the harshest months of the winter when there's not any green grass left on the ground. If you live somewhere that is very dry and has very little grass for much of the year, you should really think twice about having animals at all. Raising animals where they can't naturally forage for themselves is not sustainable. But animals can be made to create healthy pastures if you employ proper rotation techniques. The books of farmer Joel Salatin describe a sustainable grass-fed animal rotation system. Even in areas that do have good grass, we know of times when drought conditions have required farmers to travel several states away to find hay. The amount of on-farm grass and feed you can produce is a big factor in the cost of keeping animals.

Access to clean, fresh water at all times is very important. Some animals can share water sources. For instance, our donkeys and llamas share a tank and our house chickens, dogs, and cats all drink out of the same water bowls scattered around the yard. Make sure that the water doesn't freeze in the winter time. We usually carry buckets of hot water from the house to pour into the frozen bowls and tank to break up the ice in the winter. But electric water heaters are inexpensive and convenient too.

Basic veterinary care—Most large animals should be seen by a veterinarian once a year. For hardy animals like donkeys and longhorn cattle, you might get away with never needing a vet unless they have a problem. At a minimum, you should have your animal seen once by a vet to establish the relationship and to get their professional advice. Then you can decide for yourself how often you'd like your animals to be seen and evaluated. There's some debate as to how many vaccinations are needed for all the various farm animals. We tend to believe that most farm animals are over-vaccinated. The recommendations of your local agricultural extension agent are very much in line with their conventional recommendations for raising plants; they rely heavily on chemicals. But this will have to be a judgment call on your part. Some people suggest a West Nile vaccine for donkeys and llamas, but a few vets we've talked to have explained that donkeys seem to be less susceptible to this virus than horses. They suggest that the animal would have to have another pre-existing condition that weakened their immunity before it could be a problem. Tetanus shots are another issue. There's no clear evidence that indicates how often farm animals need to be vaccinated against tetanus. We've only had our donkeys vaccinated twice in the 8 years we've owned them. But we do have our llamas vaccinated every year with whatever the vet recommends.

Companionship—The more attention you give your animals, the better. They like it. They become more tame and easier to handle. You can spot any problems early. And animals provide good, old-fashioned entertainment.

Preparing for Animals

Keeping in mind the basic needs of farm animals, you'll need to prepare for their arrival. If you have animals already, you'll want to make sure you have a separate area for the new animals to go until they get settled. It's best to have a small

pen next to the field where other animals are so that everyone can get acquainted over a fence for a day or two before intermingling. We use our dog pen (without the dogs in it, of course). A temporary fence of posts and ropes is okay, if that's all you have.

Even if you stick with the low-maintenance animals as we suggest and you have good grass, it's best to have some feed for the first few days. It's a good way to break the ice and to let the animals know that you are the provider. They can usually use a little more high-calorie and protein food if they're stressed from being moved to new place. Buy some good grain or sweet feed, but don't completely spoil them.

Walk your fields and fences before they arrive and replace any fencing that might be dangerous. If you've got barbed wire from having cows previously, you should really replace that if you are introducing horses, donkeys, or llamas. We spent a whole day gathering up old rusty barbed wire that had been left around our fields before we brought in our llamas. We then put them in our dog pen for the first night. But we didn't fully check the dog pen. There were some wires sticking out that were used to attach the woven wire to the posts. Within the first four hours, one of the llamas was spooked and caught his lip on the wire. It ripped it wide open and we had to call the vet out at 9 p.m. on a Sunday to stitch him up—on his very first day!

You should also make sure you have proper halters and leads for the animals before they arrive.

CHAPTER 9

Small Creatures for the Hobby Farm

While it's good to start small in everything you do on your hobby farm, don't think that caring for small animals is somehow less labor-intensive or less expensive than caring for large farm animals. We spend more time and money on our dogs and cats than any other living creatures on our farm, besides ourselves. But small animals can be an integral part of your farm system. Chickens are almost a necessity, if you'd like to improve your pastures for larger animals or enjoy a low-cost, low-maintenance source of animal protein. Bees are invaluable in pollinating the fruits, vegetables, and flowers on your farm. Dogs and cats are your primary form of pest control. They all have their place.

Chickens

Where to Find Chickens and What Kind to Buy

Most people order their chicks through the mail from a reputable hatchery. The balls of fluff show up in a box and you put them under a heat lamp and keep them safe until they are ready to be let out into the coop. Most of the hatcheries are very good at preventing the spread of diseases and they will even inoculate the chicks against common diseases (although you may not be able to call them organic, depending on what they are inoculated with and when). But you're still contributing to factory farming if you go down this path. Hatcheries are huge operations that churn out chicks by the tens of thousands.

We suggest that you find a local farmer who sells eggs at the farmer's market. Ask around who has the best eggs consistently and who raises their chickens humanely. You may have to wait a little while until they are hatching chicks. When you find a farmer willing to sell to you, visit the farm and see how the chickens are cared for. Chickens with a lot of room to roam and clean living conditions, without trimmed beaks or wings, will produce healthy, disease-resistant chicks. Or you may get lucky, as we did recently, and find a farmer who wants to sell you her entire chicken outfit as a turn-key operation. That way you've not contributed to the breeding of yet more farm animals. It's always best to have patience in finding animals and they usually end up finding you.

A note about antibiotics and beak trimming (or docking):

Rob Rahm, of Forrest Green Farm (Louisa, Virginia) is the most productive small egg producer we know. He maintains two flocks of about 100 birds in mobile chicken coops. He also added a greenhouse roosting area that is pulled behind the main coop. This coop stays warm for the chickens and he was able to produce almost as many eggs through the winter as he has in the summer, proving that you need not use artificial lighting and confinement to keep your chickens laying all year (except in very northern climates of the country). In order to maintain the optimum health of the chickens to keep them so productive, Rob warns against ordering chicks that have been given antibiotics or have had their beaks trimmed. This is the standard procedure at hatcheries and you must make a special request that their beaks not be trimmed and they not be given antibiotics when you order. The argument (which makes a lot of sense to us) is that if you give chicks antibiotics very early in their development, then you weaken their immune system. When you get them home and stop giving them the antibiotics, they haven't built up a natural immunity to dangers in the environment and thus are more susceptible to becoming sick. Chickens with trimmed beaks (this is usually done to prevent them from eating each other's eggs) cannot forage for themselves properly and cannot catch bugs as well, also affecting their health and egg quality. You're just going to end up spending more on feed if you have chickens with trimmed beaks. We've found that chickens that are given enough room to roam and forage don't eat eggs and rarely get sick. So don't give the antibiotics, don't trim their beaks, and make sure not to overcrowd them.

The other consideration is which breed to buy. There are literally hundreds to choose from. But if you'd like to keep your chickens around for the long run as

layers, then you want a laying breed and not one raised for meat. Also, there are some breeds that are more aggressive, like Rhode Island Reds, that have terrific eggs but will do real damage to your flower beds, and the roosters can be quite ornery. We've found that almost all white chickens and roosters are docile and easier to put up with.

Chicken Advice for the Hobby Farmer

- You don't need a rooster. Hens will lay eggs with or without a rooster. So unless you want to breed your chickens, roosters are more trouble than they're worth (we've both got scars from rooster spurs to prove it). They are handsome, though. And there's something to be said about the comfort of that alarm clock each morning (roosters crow all day, not only in the morning). Never turn your back on an aggressive rooster. They will attack you and you will bleed as a result. Roosters, by our observation, do not feel pain when in defensive or offensive mode and have absolutely no fear when they become aggressive. Another reason to do without.

- Never let your chickens hatch their eggs unless you are absolutely sure you want to deal with the consequences: You will end up with a bunch of roosters that you'll need to either be prepared to eat or find someone else that wants a rooster, because roosters do not live well together (there are rare occasions when they do) and will fight each other to the death eventually. The flock we inherited recently has about eight cockerels beginning to fight each other. We've found another farmer who's interested in making them into soup and that's where they will end up. We hate to do this, but since we inherited these animals, we have to deal with them properly. It's either soup or a bloody fight to the death until the last one is standing.

- Chickens will eat anything. Keep hazardous materials away from them. We've witnessed our chickens eating paint off the side of the house (thank goodness it wasn't lead paint). We even witnessed a chicken steal a dead mouse from one of our cats and swallow it whole. A farmer friend of ours, who hunts deer in the winter, feeds the leftover carcasses to his chickens. He says it looks like someone's sandblasted a skeleton when they're done.

- If you want top-quality eggs, your chickens need a lot of room to roam. Our house flock (never more than about a dozen) has two and a half acres of fenced area to graze. So the eggs of our house chickens are richer in color and have firmer yolks than even our grass-fed chicken-tractor birds. Chickens typically lay an egg a day. If you want eggs only for your family, two or three chickens could be plenty. Eggs can last six weeks or more in the refrigerator.

- You can expect about 75 to 80 percent of your chickens to produce one egg a day in their peak during their first and second years. This drops to about 50 percent in the winter and can go as low as 20 percent when they molt. When chickens molt, they lose and replace their feathers and much of their energy is spent on feather production. You can help offset this, the same effect that cold weather has on them, by feeding them higher-protein food during their molt.
- In most cases, the color of chickens' ear lobes (the round spot straight back from each eye) will determine egg color. Brown or red lobe = brown egg. White lobe = white egg. But this isn't a hard and fast rule, especially with green eggs.
- If you don't want your chickens scratching around your yard or you can't keep your dogs from killing them, then go with a chicken tractor or mobile coop (see photos). This is by far the best setup if you want to get a steady stream of eggs, keep the chickens contained but not confined, and renovate the grass in your fields.
- Chickens get sick and chickens die. Sometimes it can be very messy and sad. Be prepared for it. There are a million diseases and health issues to which chickens are susceptible. You'll tie yourself in knots trying to diagnose them from books (a near impossibility, we've found). And there are few vets even willing to look at a chicken. The only one in our area charges $80 just for the initial exam. So if you have chickens, be prepared to put them down to keep them from suffering. We describe how to do that below.

Shelter

If your chickens are free-roaming, like our house chickens, then they don't need much of a coop except to keep them dry and free from the wind and cold. Since we have five-foot woven wire fencing enclosing two and a half acres around our house with the dogs as guardians, then we only have a small coop with a couple of roosts and a couple of laying boxes for our house flock. The chickens only spend the night in there or go in when they need to lay an egg. A house flock like ours is the

The chicken palace at Poindexter Farm comfortably houses one hundred birds.

easiest and lowest-maintenance way to keep chickens. If you're keeping your chickens confined to a smaller area, you'll need a bigger coop that offers them at least two feet of space (more is suggested) for each when they are all in the coop.

Another reason to give chickens as much room to roam as possible is that you'll have less to clean up after. A coop that has chickens in it all the time needs to be cleaned once a week at least. And chickens confined to a small yard will quickly turn it to dirt and mud (albeit nutritious dirt and mud). Our house chickens' coop typically only needs cleaning once a month. And by the time we do, most of the manure has begun to compost and it's easier and less smelly to deal with.

At night, chickens like to roost and bunch together. Chickens are easily trainable to go into a coop at night. If they don't do it on their own, wait for them to start to bed down for the night, catch them, and put them in the coop. Usually it only takes a couple of nights for them to figure out where they should go to sleep. Once they're in at night, shut them in and latch the door. Predators are very good at finding ways into coops. In the morning, open the door to let them out.

Chicken Tractors and Mobile Coops

Chicken tractors (which are movable chicken coops) are the way to go if you want more birds than a regular yard will handle, if you want to produce eggs for sale, or if you're interested in keeping your pastures healthy for other grazing animals. The coop is moved every few days with a tractor, truck, or lawn tractor

and chain in order to give the chickens fresh grass and to spread their manure over the fields as a natural fertilizer rich in nitrogen. You can leave them on one spot longer to really give an area a natural boost. Parking a chicken tractor over your garden in the winter is a great way to boost your soil. Just be careful you don't overdo it and end up giving your garden a nitrogen imbalance.

We only advocate chicken tractors that have an electric fence enclosure to give

RIGHT: Our chicken tractor surrounded by an electric fence. The chickens will be moved around the garden to fertilize and eat harmful insects.

chickens lots of room to roam and forage. The popular chicken tractors that are low to the ground and fully enclosed are just glorified cages. Sure, the chickens get fresh grass every few days. But chickens like to roam, peck around, flap their wings, and run after bugs. They like to roost above the ground at night. And it shows in the color and richness of the eggs. The more confined chickens are, the more you will spend on off-farm food to feed them. We have a chicken tractor we bought from a farm that was shutting down. It's the kind described above, but we use an eighty-foot electric fence around it to give them room to roam. But it's difficult to move with the chickens in the coop, and then it becomes a trick to get them inside the fence, move the coop, then get the fence back up without them escaping. When we've tried to move the coop with the chickens inside, they become frightened and have been caught under the coop as it's being moved (never being seriously injured though). So it's not the most convenient way to keep chickens, although it is very good for fertilizing the ground. You should move them every few days, depending on how fast they are eating the grass and how much grass you have to begin with.

A mobile chicken coop on wheels or that's raised above the ground is, in our opinion, the best method. If you're using an electric fence (solar electric is quite convenient), just wait for the chickens to bed down in the early evening, close them in the coop where they are roosting comfortably, roll up the fence (make sure to turn it off first), move the coop to fresh grass, and set the fence back up. There's no drama of frightened chickens running this way and that. This can even be adopted as a smaller version for your yard without the electric fencing, as long as you have a door to shut them in or a dog to keep guard at night.

Enclosed Coops with Chicken Yards

This is the least economical option for raising chickens and produces eggs only as good as the grain you feed the chickens. But if you live in an area that has little grass for forage or somewhere that prevents you from rotating your flock or

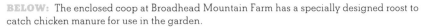

BELOW: The enclosed coop at Broadhead Mountain Farm has a specially designed roost to catch chicken manure for use in the garden.

The previous owners of our farm left a rooster and hen behind when they moved because they couldn't catch them. Ted was a handsome white leghorn rooster, with a floppy red comb. Beverly was his svelte female companion, who rarely left his side. Ted and Bev had both been spared the injustice of wing clipping and beak trimming, so their features were fully formed and striking. Ted's spurs were formidable; at least four inches long and they curved to sharp points, though they served only as a warning and were never used in anger.

Ted was a true Southern gentleman. His barrel chest bulged with pride as he strutted around his farm. He paid particular attention to Beverly, always scanning for trouble and calling to her if she were out of his keen sight. He would crouch protectively over Bev when she was laying her daily egg so he'd be the first to confront anyone who dared interrupt.

Ted's crow was less impressive. It sounded less like a cock-a-doodle-doo and more like a gurgled Rebel yell. It could be described as a cross between a war whoop and a rabbit scream made by a sick goose. But what it lacked in sonorous melody was made up with the enthusiasm of delivery, which could come at any hour from before dawn to dusk. Bev, besides enjoying the body of a poultry supermodel, could easily fly to the tops of the two towering holly trees in the front and back yards of the farmhouse, where she perched and slept each night. Ted, too top heavy to fly, would settle down on the front porch steps to await the dawn.

While we loved the idea of Ted and Bev and the novelty of a daily egg hunt, neither of us knew much about chickens. Ted and Bev had no coop to sleep in at night as the previous owners just relied on their dogs to keep predators at bay. They hadn't even constructed a laying box for Bev, so she used a flower box outside our living room window as a substitute. We briefly fed them cat food until discovering that most cat food has chicken in it, and we just didn't feel right about that. Bev was able to take comfort at night in a tree, but Ted slept on the front steps, exposed to the elements. We figured that they had been getting along just fine that way before, so we weren't in any hurry to improve their housing situation.

It wasn't long after we moved in that we awoke one morning, concerned because Ted had not begun crowing and the sun was now fully above the tree line. We quickly went out for a look, only to find a trail of white feathers leading from the porch, across the field, and into the woods. Ted had no doubt met his demise in the mouth of a fox. And thus we learned our first lesson of hobby farming—chickens and other defense-

less farm animals will die if not protected at night.

Our guilt at not protecting Ted properly weighed on us. Bev had been laying eggs in the window box. So we let her become broody in hopes that she might hatch out Ted, Jr. Audrey had the idea of memorializing Ted by naming our farm in his honor. This is how Ted's Last Stand Farm & Gardens came to be.

Three weeks later, Bev hatched out nine fluffy chicks in the window box. By that time, we had built a coop to keep them safe. Soon after, when seven of the nine chicks turned out to be Ted Juniors and not Bev Juniors, who crowed incessantly and fought each other to bloody pulps, we realized our second lesson of hobby farming—unless you're prepared to live with the cacophony of multiple roosters, find homes or cooking pots for them all, or kill the males yourself, then don't ever allow your chickens' eggs to hatch. And so, we found another farmer who enjoyed making homemade chicken soup and we've felt guilty about that ever since.

ABOVE: Ted.

letting them roam free (like the mountains where there might be lots of predators), then this might be your only option. But be careful not to overcrowd them or they will resort to egg eating and cannibalism.

Food and Water

Chickens will eat anything. But *anything* gets into the eggs you eat. So be picky about what you feed them. The regular chicken feed and scratch at the feed store has pesticides and other chemicals you probably want to keep out of your eggs. Our local feed store now sells organic grain, but it's quite expen-

RIGHT: Water and food are placed near the door for easy access.

LEFT: A solar electric fence makes this mobile chicken coop easy to move anywhere on the farm.

sive. After some time looking, we've now found a local chicken feed producer that sells non-GMO food without chemicals. But it's not certified organic. Our customers at the market don't seem to care about an organic label; they only want our word that we're not giving them any chemicals or antibiotics in their food. If your chickens have as much room to roam as our house chickens, then you don't even need to feed them except in the winter when there's no green grass. Given the choice, chickens eat mostly grass and bugs (more bugs result in deeper colored yolks). So in the winter, you may need to supplement their grazing. The chickens in our chicken tractor eat more food than our house flock, but we try to keep it to a minimum and force them to eat the grass instead. There's no formula if you're keeping mostly grass-fed chickens. You just need to pay attention to the amount of grass and grain being eaten and the production of eggs.

We combine equal parts organic chicken scratch with the non-GMO layer feed and add oyster shells (the bag will tell you the proper ratio) for calcium and grit to help them digest their food. You can also just leave out oyster shell or grit in a bowl for them to feed on when necessary. If you only have a house flock, just throw the food on the ground and let them forage. Only throw enough for them to feed for about five minutes and you can do it in the morning and the evening. If you're keeping them in a smaller area or in a chicken tractor, then you'll need a feeder in the coop for them. If they have plenty of grass, they shouldn't eat more than a quarter pound of feed a day per chicken. We've found that chickens also like table scraps, breadcrumbs, the crushed chips at the bottom of the bag, and especially grapes. They love grapes. Meat has pathogens, so don't feed it to chickens. If you've got a chicken tractor operation, then you'll need a little trial and error to figure out how much feed to give them. It all depends on the size of the flock and the size of the area they are foraging. You want to encourage them to forage, so don't let them fool you into thinking they'll only eat grain.

Chickens need water at all times. Dehydration will kill a chicken very quickly. A droopy, dull comb is a classic sign of dehydration. For our house flock, we keep dog bowls filled around the yard and all the animals use them. For our chicken tractor, we have one large watering can that we fill each morning and hang from a chain connected to the top of the coop. You might consider a water heater if you live in a place where it stays really cold for long periods. During the coldest part of our winter, we just fill a couple of five-gallon buckets with really hot water and go around and pour it into all the bowls. This typically unfreezes them and cools the hot water off enough for the animals to drink. Chicken tractors and coops need both grain and water feeders at all times.

RIGHT: A chicken house built on a hay trailer is easy to move and sturdy enough to carry large barrels of feed and water.

Can You Profit from Your Chickens' Eggs?

Let's begin with a quick calculation of the costs. Perhaps you got lucky and found a used chicken tractor or mobile coop already built and you got a great deal. But you had to buy an electric fence with a solar charger, by far the easiest to use for a pasture-fed operation. You bought fifty chickens close to laying age in the spring. Add in some feeders, water troughs, and various food scoops. You'll need about three bags of supplemental feed each week and some oyster shell or grit. You'll need several sealed plastic containers to store feed. Egg cartons are 25 cents apiece on a good day, although you might find that customers bring you used cartons once you get established. So using very conservative estimates:

The chickens will start laying gradually and will finally reach their peak at the beginning of the summer. Say you have a really productive flock and you get forty-eight eggs a day out of fifty hens (this is a rosy scenario; 75 percent egg production is more typical). That's four dozen a day, seven days a week (chickens don't rest on Sundays). That comes to twenty-eight dozen a week. The average price at our market is about $3.50 a dozen, although some people get as high as $6. That's about $98 week in sales. So it takes roughly eighteen weeks to break even, if you're lucky. Chickens slow down production in the winter when it's cold and when they go into molt. Also, you might have a harder time reaching retail customers in the winter and you might have to sell them cheaper to a reseller. If for the rest of the thirty-four weeks out of the year, you average 20 dozen, that's 680 more dozen you have to sell. That would be $2,380 in profit, assuming you have a steady direct-to-customer base and you don't have to sell wholesale, which would cut that amount in half. So, you stand to make a maximum of $2,380 in your first year, raising fifty grass-fed chickens and hustling to develop a loyal customer base you can sell direct

50 chickens near laying age ($6 each) = $300
Chicken tractor (used) = $250 (new ones are $800 to $1,000)
Electric fence = $175
Solar electric charger = $185
Feeders (2 @ $18), water troughs (2 @ $20), and metal scoops ($24) = $100
Plastic feed containers (2 @ $35) = $70
Enough of the cheapest feed and oyster shell you can buy to last the summer (five months) - $700 (this can be much higher if you go with organic)
Total = $1,780

to year round. They never keep up the same rate of laying after their second year and will steadily decline, but you also won't have the startup costs each year, so after about three years you might double your profit.

If you can do this, you'd be the most successful chicken egg hobby farmer we've met. But there are many variables in raising chickens, especially if they come down with a communicable disease. One farmer we know had his flock come down with one of the many diseases that will make chickens quit laying for good. He had to put down ninety birds himself and he didn't feel right about eating them or selling them for meat since they were diseased (most of the producers you find in the supermarket aren't so discerning). A heartbreaking task, both personally and economically.

Tom Martin of Poindexter Farm in Virginia buys chickens just as they're ready to lay and sells them at the end of each farmer's market season to friends and neighbors that want laying hens so he doesn't have to pay to feed them through the winter when they're less productive. In this way, he's a bit more profitable than most small chicken farmers. But as you can see, it's very hard to get rich selling eggs. And raising chickens for meat is not that different than raising them for eggs. Our economic formula would only need slight tweaking for meat birds and your potential profit would be about the same. Your labor would certainly go up, though, as you'd need to kill and dress them. Unless you're going to grow exponentially and make the jump to becoming a factory chicken farmer, it's not the way to riches.

Money isn't the only benefit you get from raising chickens, however. Knowing that your eggs come from chickens that eat all-natural foods and aren't given any antibiotics or chemical-laden feed is worth a good deal to most people. Chickens also create valuable fertilizer for your garden and renovate your fields without chemicals. They eat lots of pests as well. They are one piece in a big puzzle of diversity on your farm. So like all the hobby farming ventures we promote in this book, don't do it to get rich. Do it to enrich your life and your farm first and if there's a small profit from it down the line, then all the better.

When Something Goes Wrong

Chickens are tough creatures physically, but they are very susceptible to disease. And it's very difficult to diagnose a chicken unless you've had a ton of experience and worked with someone who knows all about chicken problems. Books will drive you mad like online medical sites do. You'll find yourself believing all your chickens have avian flu by the time you put the book down. Chickens get parasites, frostbite, bacterial problems, and coccidiosis (coxy), which is probably the most common chicken disease. They can also have egg binding, which is when

Apple Cider Vinegar

Apple cider vinegar is an all-natural supplement that you can add to your chickens' water in a quarter-cup per gallon ratio. It has vitamins, minerals, and trace elements. It lowers the pH in chickens' stomachs to help with digestion. It's also an antiseptic, which helps kill germs and boost a bird's immunity. But you'll want to give your chickens the raw, unfiltered kind so that they get all the benefits. Most health-food stores or Whole Foods carry it. And you might find it in your local feed store or co-op. Also, only use plastic water containers, as the vinegar will corrode metal. Some farmers give it as a regular supplement to ward off disease. Others wait until they see a problem with their chickens before using it. The science isn't in on which option is best, so a little trial and error is the best course of action.

an egg gets stuck coming out. This is one of the easier problems to fix and involves a little Vaseline and a rubber glove.

You'll know when there's a problem. A sick chicken will stay away from the rest of the flock, lower its tail, puff up its feathers, and not move or graze much. If you spot a problem, isolate the chicken from the others. You do need a separate cage for this purpose. A dog crate works well. Make sure the chicken has food and water, and put electrolytes in the water. That's about all you can do unless you find someone who knows something about chickens that can help. Most of the time, the chicken will perk back up soon.

But sometimes chickens get sick and don't get better. They get weak with diarrhea and become dehydrated. Sometimes you'll just find them dead. Other times, they have a serious problem and it's obvious they are suffering. We had two roosters for a time. They got along well enough until one of them started showing signs of age. The younger one would periodically beat up the older one (you've probably heard of the pecking order) and we'd have to separate them. One day, the younger rooster really let the old guy have it and wounded him to the point that he couldn't walk. So we had to put him down.

When a Chicken Meets Its End

We've had to put other chickens down since then too. This is the easiest way to do it that seems the least traumatic for chicken and human:

- First and foremost, be calm and deliberate. Suck it up. It's your responsibility to keep the animal's suffering to a bare minimum. Prolonging that suffering because of your own emotional frailty is unacceptable.

- Hold the chicken upside down for a couple of minutes by its legs. This allows the blood to rush to its head and makes it calm.
- With its comb facing away from you, pin its head on the ground with a broomstick or something similar.
- The stick should be right behind the head and you hold the stick down on either side of the chicken's pinned head with your feet.
- Now, with the chicken's head pinned between your legs, pull up solidly on the chicken's feet so that its head and neck stretch. This separates the head from the spine.
- Hold that pressure (not too much or you'll have a headless chicken, but that works too) until the chicken stops flapping its wings and dies. This is a nervous system response and doesn't seem to create any suffering.
- Dig a deep hole where your dogs can't get to it, say a few words of praise, and bury it.

Honeybees

After moving out to rural Virginia, getting bees was one of those things we often thought about doing. Just as we had obtained our donkeys, then our llamas, and built our pond, bees were a part of the vision of our farm. After establishing our large cutting garden, it began to seem that we were lacking a beehive rather than just hankering after one. So, Audrey finally enrolled in a beginning beekeeping course offered by our local agricultural extension agency. The class provided much valuable basic information and prepared us for what to expect from a new hive, as well as describing in detail what we would need in order to get started. The reference materials have been invaluable, as has the knowledge that help is just a call away. Taking the class not only made us feel comfortable about keeping bees, but also piqued our enthusiasm. We love to watch the bees filling their pollen sacks, always trying to get more pollen stuffed in to carry back to the hive. As we cut flowers, weed, and tend to our vegetables, the worker bees leave us to our tasks; they are far too busy with their own jobs to pay us any heed.

Last May we got our first hive. It was a colony living in our friend's wall. An expert from the local beekeeper's association came out and helped to remove the bees, along with a section of their comb, and trapped them in a bucket for transport to our farm, which is about twenty minutes away. For his troubles the master beekeeper got to keep all of the honey those bees had stored up. The bees adapted to their new home quickly, it seemed. But soon we realized there was a problem—the queen hadn't made the trip. So we ordered a new, Russian queen, and inserted her into the

Audrey getting ready to open the hive.

Removing the inner cover.

The smoker makes the bees retreat into the heart of the hive.

Feeding the bees with sugar syrup.

hive. Not long after that, they swarmed and were gone with the wind.

Our beekeeper friend, Harold, quickly brought over two well-established hives as a replacement. Beekeepers are always looking for convenient places to put new hives. This is an easy way to get into beekeeping without taking on all the expense up front. Befriend a beekeeper who will put a hive or two on your place. Learn to take care of it; benefit from the pollination they provide; then get a share of the honey at harvest time. It's a can't-lose situation.

Many people are enticed into beekeeping by the lure of the delicious honey the bees produce. However, a honeybee's main, and most valuable, purpose is pollination of crops. A healthy, sustainable honeybee population is essential to producing sufficient yield and quality of many vegetable, fruit, and legume crops for human and animal consumption. Keeping honeybees is also a great family project to explore and learn about the life and importance of the bee. They are an integral part of a sustainable hobby farm. While you might do without them, your plants will thank you if you introduce bees into their world.

The recent recognition of colony collapse disorder in beehives has seen a sometimes dramatic decrease in bee populations around the United States and other Western countries. Beekeepers have seen a sudden die-off of their hives or an unexplainable disappearance of the bees. The reasons for this disorder are still unknown, but culprits might include pesticides, global warming, fungal infection, or a virus. This is more reason, not less, for jumping in and keeping bees. The cause of colony collapse is unknown, but adding healthy hives to small and diverse farms that shun pesticides and other chemicals can only be a positive development in helping the honeybee to survive.

As with the other endeavors discussed in this book, beekeeping, or apiculture, is an enterprise that will grow as your interest in it deepens. Most first-time beekeepers order a package of honeybees and a basic starter kit. The first season is one of hands-on learning. If you find you enjoy the time you spend handling the bees, become fascinated by their behavior, and are undaunted by setbacks, you will most likely expand your beekeeping operation. Often, however, the amount of time, attention, and money needed to keep the honeybee colony active and healthy is more than anticipated. Just as with gardening, your time and care are the essential factors to success.

Getting Started

Learning About Beekeeping

It's strongly recommended that you take a beginning beekeeping class or workshop before setting out on your own. These courses are most often offered in late winter or early spring (February and March), the time when beekeepers are starting to see activity in their hives after the winter, and are making plans for their bees for the coming season. A course not only provides valuable information, but provides a multitude of resources through meeting other beekeepers in your area. A typical beginning beekeeping class generally covers the following topics: organization and management of the colony; the different types of bees and their jobs; the life cycle and development of the honeybee; basic bee anatomy; bee diseases, parasites, and pests; how to get your colony started; how to handle your bees; honey production and processing; and how and where to order equipment in your area. As you work with your colony of bees you will

RIGHT: This beekeeper is beating a hasty retreat after tempting fate by opening the hive without his bee suit on.

undoubtedly have questions; knowing people nearby who can help out is a great advantage. It is also worth the time to peruse several beekeeping books and magazines. Taking a beekeeping class and talking with other people who are interested in and/or keep honeybees will help you to decide if this is an enterprise in which you want to invest.

Stings

Two things pop into most peoples' minds when the subject of bees is mentioned. These are honey, and getting stung. Neither of these should propel you into, or away from, the pursuit of beekeeping. In general, most people have an aversion to being stung. Honeybees sting to defend themselves or their colony. By using proper handling techniques most beekeepers are able to minimize the occurrence of bee stings and people typically build up immunity to the stings the more they're stung. If you keep bees you will most likely be stung at some point in the active bee season. Those people who have a strong allergic reaction to bee stings should avoid any contact with bees. Also, people with very strong allergies to pollen will want to avoid the beehive. If you or someone else is stung, get away from the bees as quickly as possible and remove the stinger. If the person who has been stung has an allergic reaction, get medical help immediately.

Beekeeping Equipment and Costs

As with any new project or hobby, beekeeping requires an investment in the proper equipment and basic starting materials. In general, the following is a list of what you need to get started keeping bees. Costs for these items will vary by company. A list of beekeeping equipment suppliers is provided in the appendix.

Bee Keeping Essentials

- ✓ Bees, package or nucleus, including queen
- ✓ Brood boxes with frames and foundation
- ✓ Bottom board, outer cover, and inner cover
- ✓ Honey supers with frames and foundation
- ✓ Protective clothing
- ✓ Hive tool
- ✓ Smoker and smoker fuel
- ✓ Sugar
- ✓ Feeder

The Bees

In order to get a hive up and producing its first season, you will need to order bees from a reputable source in January or February. You can order bees in packages of two, three, or five pounds. Each pound represents approximately 3,500 bees. You can also order bees in a nucleus colony. A nucleus colony is a hive containing bees in all stages of development. The nucleus (or nuc) gives you a jumpstart on honey production because its queen is actively laying. The frame has comb already built up on the frames (called "drawn comb") with brood present in these cells. With a package of bees the workers must spend time and energy building up the wax on the frames in the hive before the queen can begin laying.

It is important to realize that the first year of a bee colony is a building year. What this means is that you will not be extracting any honey from the hive this first year. With luck, your colony will produce enough honey to feed itself over the winter months. It is more likely that you will have to do some supplemental feeding in the fall and spring by dissolving sugar in water and pouring it into the bees' feeder.

The Hive

After you have ordered your bees, you will need to order your hive. There are several different configurations, but these are the general components you will need for a one-hive beginning beekeeping adventure.

A hive stand is necessary to elevate the floor of the hive off the ground. This does not have to be a piece of purchased equipment. Your hive can rest on concrete blocks, logs, bricks, etc.

A bottom board, which is the floor of the colony and the main takeoff and landing point for the workers. Screened bottom boards can aid ventilation, as well as allow the detritus of the hive to fall to the ground, thus helping to prevent disease and mite infestation.

The hive bodies, which house the brood and honey and are the outer structure of the hive. The hive bodies come in different sizes that allow beekeepers to tailor their hives to their individual preferences. There are two purposes for the hive bodies—brood rearing and honey storage. The bottom hive body is usually of deep or medium depth and is called a *brood chamber*. This is where the queen will spend most of her time laying, and where the workers feed the larvae and then cap their cells in beeswax for their final days of development. On top of the brood chamber is a hive body called a super. The super is for surplus honey collection.

RIGHT: The brooding chamber filled with frames sits atop the bottom board. Concrete blocks are used as a hive stand.

The hive bodies are filled with wooden **frames**. These frames consist of a sheet of beeswax or plastic comb foundation suspended in a wooden frame. The worker bees add wax to this foundation to make the cells of the comb that are then used for brood rearing and the storage of honey and pollen.

Your hive will need an **inner cover** that rests on the uppermost super and sits beneath the outer cover. This inner cover serves several purposes. In addition to preventing the bees from gluing down the outer cover to the super with wax and propolis (sap or resinous materials collected from trees or plants by bees and used to strengthen the comb, close up cracks, etc.; also called bee glue); it provides an insulating air space between the outer cover and the hive. It also allows for the easy attachment of a feeder. Finally, the **outer cover** protects the hive from weather. It is generally composed of wood covered by galvanized metal to prevent leaking and weather damage. The sides of this cover hang down over the hive, thus it is called a "telescoping" cover.

When your hive arrives you will need to paint the outer surfaces to protect and preserve the wood. Use white or any light colored paint in order to prevent heat buildup in the hive during sunny days.

You'll want to make sure to buy **protective clothing for yourself;** at minimum, a bee veil and gloves. You will also need a **hive tool**. This is a metal bar used for separating hive bodies, and prying apart frames in the hive bodies that have become stuck together with wax and propolis. A **smoker** is the final essential piece of equipment for working bees. Cool smoke repels the bees and reduces defensive and aggressive behavior. The bees react to the smoke, believing there is a fire, and they retreat into the hive. A smoker allows you a measure of control over bee behavior. Hopefully, knowing this will help you maintain a calm and relaxed manner while working with your bees.

To get your new colony of bees off to a strong and healthy start you will need to provide them with food in the form of sugar syrup. This allows the bees to concentrate on building up the wax comb rather than having to forage for nectar to survive. It is likely that you will also need to feed your bees in the fall and winter to ensure that they have enough food to survive until they can begin to forage again in the spring. There are different types of feeders available through bee suppliers, and most beekeeping books provide instructions for making your own feeder.

Once your hive begins to produce surplus honey you will need to invest in extraction equipment. This can be quite expensive for the hobby beekeeper.

There may be options for renting or borrowing equipment in your area through a local beekeeping association. Other beekeepers with extractors are usually willing to do it for you for a 15 percent share of the honey.

Dogs and Cats—The Heart of the Farm

We were idling down our gravel road when we spotted a mangy dog just sitting there alongside it, taking a load off after a long walk. He was skinny, maybe twenty pounds, a blue heeler mix, and had that nervous look of a stray in his eyes. He was black, white, and brown spotted, as if he'd been sitting by the side of a puddle when a truck drove by. One of his ears was mangled and flopped over while the other stood straight up.

We already had three dogs, one of which, Casey, was a recent stray that had followed our dog Tess home from one of her walkabouts. Tess was always going on walkabouts, visiting all the dogs in the neighborhood, many of which followed her home to our house in hopes of a date or a good meal. And living in the country, it's certainly not uncommon for dogs to show up unattended. But we'd recently tried to put a stop to that.

Our first dog, Tippy, had recently followed Tess on a walkabout and was hit by a car up on the main road. We were out in the yard one morning watering the garden and we spotted Tippy just sitting down in the field. I called him, but he wouldn't come. I went out to get him and realized he was bleeding and not able to walk. He'd broken his pelvis, but still made it a good 200 yards towards home before he couldn't go anymore. I scooped him up and we sped him to the vet where he was fixed up for a mere $2,500.

After that, I decided a fence around our property to keep our dogs in and other dogs out would give us some peace of mind and would be cheaper than advanced orthopedic veterinary surgery. It wasn't long after I got the fence up, a mostly woven wire and T-post construction encompassing three acres, accented with some regular wood fencing and a shiny red gate for our driveway, that we spotted that mangy dog.

I pulled up alongside him, rolled down the window of the car, and told him in no uncertain terms, "Go on now, dog! Get on home!" He turned with his tail between his legs and looked back at us, head hung low, and hobbled a little further from the road. We hoped he'd go on his way, but he'd only moved a little closer to the house when we spotted him upon our return.

I made another show of ordering him away and then pulled our car into the gate and secured it behind me. He now stood in the middle of the road, looking right at me. But I wasn't falling for it. I told Audrey and Mary, my mother-in-law who was visiting at the time, "Ignore that damned beast. And whatever you do, don't feed it. Surely he'll be gone in the morning."

So we kept the other dogs in the house, especially Tess, so as not to encourage him. The next morning, I went out to the gate and there he was, this time wagging his tail and pacing about, like he really expected me to let him in now that he showed how much persistence he had. And right beside the gate, there was a small trench where he'd obviously spent most of the night digging. Here was a dog who was actually trying to dig into a fenced area. What the hell was wrong with this dog?

I couldn't keep him out any longer. I let him in and he quickly won the other dogs over. Mary took out a full can of wet dog food and he ate it in about two bites. Then he immediately flopped onto his back and rubbed it all over the grass in pure joy, like an astronaut or hostage kissing the ground after returning home from a long ordeal.

So we kept him. And we called him Jack. He immediately took over the house and claimed the couch as his own. Soon it was like he was allowing us to live in our house and not the other way around. He'd herd the other dogs mercilessly and he growled constantly at every movement like an old man grumbling under his breath at the idiocy of the world. He was a loveable curmudgeon.

About six months had gone by when we had an unexpected visit from the previous owner of our farm. She was in the area and just wanted to stop in to see what we'd done with the place. I was just wrapping up the full tour when she spotted Jack and stopped cold in her tracks.

"Danny, is that you?"

Jack slowly walked over with his head hung low.

"That is Danny!" she exclaimed. "He's got that mark on his back and that ear that's crooked! We thought you were dead."

Turns out, Jack was really Danny and he'd belonged to the previous owners. It immediately dawned on me that I'd actually met Jack when I was looking at the place before we'd bought it. Once the previous family and Jack (then Danny) had moved away, they had lived all over the area, in three different spots over four years. One was over thirty miles away. Recently, they had moved about ten miles as the crow flies and Jack had soon disappeared. They assumed he'd run off and been hit by a car or shot. But what had really happened is that Jack realized how close he was to his real home and he'd set out on a great adventure as the prodigal son.

Considering all the effort he'd made getting back, the woman certainly wasn't going to force him to leave again. Later, I rummaged through

ABOVE: Jack

some old photos of the house that I'd taken when we were house shopping.
Sure enough, right there in one of the photos of the front of the house was
Jack, standing on the edge of the porch like he owned it and taking stock of
this strange guy pointing a camera at him. Little did we both know at the
time that we'd become lasting friends and would spend Jack's retirement
years together in his boyhood home.

—Michael

This story isn't uncommon in the country. If you wait long enough, dogs and
cats will show up at your farm on their own. If you think the city has a problem
with stray animals, you haven't been in the country long. Our friends Harold and
Mary Plasterer founded and have helped run a local spay and neuter program. They
collect donations and grants to pay for spaying and neutering of dogs and cats.
They are regularly called out to farms where there are thirty or more feral cats.
This can happen very quickly if an owner is not paying attention. Harold and Mary
use humane animal traps to collect all the cats (you can't even leave one) and take
them to veterinarian clinics that provide low-cost spaying and neutering. They
then return the animals to where they were trapped and release them. They've
prevented tens of thousands of unwanted cat and dog births. If you can't physically
participate in these types of organizations in your farm community, you should
consider donating and/or adopting from them. The animals just over your fence
are still a part of your farm ecosystem.

Just because you live out in the country doesn't mean your dogs shuld be free
to roam. Dogs get into all kinds of trouble in the country and many or hit by a car.
After one of our dogs was hit by a car, we installed a five-foot woven wire purchase
we've ever made. Nowadays you can buy an inexpensive "invisible fence" that
creates a barrier wherever you choose to set it and an electric collar on your dog
keeps them inside of it. We can't recommended enough that you keep your animals
confined to your own property.

Dogs and cats go hand-in-hand with your hobby farm. They are your constant
companions, riding shotgun to the dump
and guarding you against invaders. You'll
quickly be overrun with rodents if you
live in the country and you don't keep cats.
Cats enjoy living outside and can be almost
maintenance free this way. There are just
too many country animals that need a home.
Adopt them, spay or neuter them, and share
their happiness at being so lucky as to end
up on a happy, healthy farm.

Can't We All Just Get Along? Yes!

Most of the animals you'll keep on your farm are natural enemies—dogs and cats, dogs and chickens, dogs and horses. Well, we guess it's really the dogs that are the problem. And there certainly are dogs that have that taste for chicken that you may not be able to cure. But we've found that if you slowly introduce your animals to each other, while supervising at all times, they will learn to get along. But the most important precursor to this is for you to establish your rightful spot as the pack leader. If your animals do not respect you, then they certainly won't have any reason to respect other animals in the pack.

Always keep animals separate initially when bringing them onto the farm. Keep them in a cage or pen that the other animals can approach and smell. Then introduce your dogs one by one to the new animals (you don't want them to get the pack mentality if they're all together) after a couple of days. Cradle the cat or chicken and get on the ground with the dog (best to do this inside, if it's a cat). Call the dog over and allow him to sniff. He will probably try to nip at the animal and that's when you give him a sharp rebuke to tell him what's appropriate. Then call him over again. When he gets that predator look in his eyes and lowers his head, sharply call his name to get his attention off the other animal. Keep eye contact with him and tell him what you expect. He won't understand the language, but he'll understand the tone. The most important part of this exchange is to get the dog's attention off the new animal and to direct it to your eyes so he knows you are completely focused on him. Spend time around all the animals together and make sure they know that you expect them to accept the newcomer into the pack. And whenever the dog seems to fixate on the other animals, sharply call his name and make eye contact with him again. If a dog will not ever completely calm down, our method of fixing this problem is to grab him by the scruff of the neck and look him directly in the eyes and assertively tell him, "No." Don't yell it. You should be very careful doing this, however. If the dog does not already respect you as the authority, he may bite you.

Sure, we've had our problems. We came home early one day to find our new rooster cowering behind a pile of wood with half his feathers ripped off. We probably didn't react the way the professionals tell you to. We picked up that rooster, ran after the dogs screaming obscenities at them, and locked them in their pen. They say that you can't correct a dog unless you catch him in the act. But our dogs knew what they had done and they never did it again. The rooster made a full recovery.

So stay around your animals when they are mixing until you are certain they have made nice. Never leave them alone together until you're absolutely sure everyone knows their boundaries. Sometimes it's just best to keep animals separate and that's what good fencing is for. Good fences make good neighbors, both for humans and animals.

CHAPTER 10

Large Grazing Animals

Our low-maintenance philosophy of keeping animals means we'll only cover those grazing animals that will be less likely to take over your life (horses are an exception to this rule). So while sheep are cute as pie and offer yearly fleece you can sell, and goats create delicious cheese and can clear land like no other animal, and dairy cows can produce milk you'll never find in a store, they all require more time than a hobby farmer will probably want to put in while still holding an off-farm job. The smaller animals require more extensive fencing and housing. They are more susceptible to sickness and predator problems. And anything that requires milking once or twice a day isn't worth the trouble, in our book. But after you've explored other opportunities on your farm and you decide that goat's milk cheese is the market that needs filling in your area, then jump right in. We just wouldn't suggest it at the beginning of your farming life. Remember, we're promoting the *joy*, not the *toil* of hobby farming.

Handling Large Animals

Temple Grandin, in her book *Animals in Translation*, describes going out into a field of skittish cows and just lying down. The cows, being naturally curious animals, soon gathered around her and quietly checked her out. Putting herself down at the level where they keep their faces most of the day was how she connected with them and convinced them she was no threat. Cesar Milan, in his show *The Dog Whisperer*, advocates a calm, assertive demeanor around animals. We take both approaches.

Animals like horses, cows, longhorns, donkeys, and llamas are powerful creatures that can kill or maim a person if startled or threatened in the slightest. They can sense fear too, like all animals. So always screw up your courage when approaching them, control your breathing, and be forceful without being cruel. Talk to them in low, reassuring tones. They don't speak baby talk, so avoid that temptation. Avoid making sudden moves around them. When walking behind them, give them a wide berth, or keep your hand running along their body so they know where you are. And once you're comfortable, then get down on their level

like Temple describes in her book. You'll be fascinated and surprised at the perspective you gain looking up at them instead of the usual standing position that has you always looking down on them. But if an animal is acting out, it's okay to give him a swat on the hindquarters to let him know the behavior isn't appropriate. These are big, tough animals and can take a firm smack on the rear. If you have a very unruly animal that you're having a hard time caring for, you should call in another more experienced farmer or veterinarian to help you find strategies for dealing with the animal. And there's one almost foolproof way to get an animal to do almost anything—find his favorite treat, like apple and oat treats or sweet feed shaken in a can.

LEFT: Rico the llama peeking out from the llama shed, where he's free to come and go as he pleases.

Shelter

Large grazing animals require little in the way of shelter. In most climates, a three-sided shelter with a roof is about all that you need to shade them from the sun or protect them from the cold, and much of the time the animals won't even use it. If you live somewhere that has frequent or even occasional heavy snowfall, then you will want to offer a shelter where animals can dry off. It's usually not the cold that's the problem, but the wet. If you plan to breed your animals, you'll definitely need a place for the newborn and mother to shelter until the newborn is ready to be put out into the field. And you'll need to plan for separate living quarters for your unfixed males.

We have a small stable for our two donkeys and two llamas. They share it and tolerate each other. It has no door, so the animals come and go as they please. It has two windows, which we open in the summer and close up for the winter. We also have an area to store hay where the animals can't get to it. When we had over fifty inches of snow this winter, the animals spent a good deal of the day in the stable, but they would come out to eat up by our house when we'd call them. The more time they spend in the stable, the more manure builds up that you'll need to muck out. The llamas could have lived outside the entire winter, had we not spoiled them! They are mountain animals, after all.

Cows and longhorns are even tougher and typically prefer to shelter under trees and not in enclosed areas. You can ask local farmers what, if any, shelter you

might need for your climate. Remember that cows live out on vast ranges in the West and Rocky Mountains and they're not given shelter. Of course, they freeze to death in these places occasionally, but that's certainly not common.

The best fencing for donkeys, llamas, and horses is board fencing or woven or high-tension wire fencing (see page 39 and 40). Some people use electric fencing, but large animals will eventually break through it or the deer will inadvertently knock it down. Electric fencing is ideal if you are employing rotational grazing. Barbed wire is strictly discouraged if you have any animals other than cattle. Cattle are bigger, tougher, and more destructive and most need to be kept in check with electric fencing or barbed wire.

Food

Large, grazing animals can live on grass alone, with minor mineral amendments. That is, if you have enough for them. The more feed you have to give them from off the farm, the less sustainable and economical is your farm system. We keep our animals mostly for the compost we create from their manure. But if we had to spend $1,000 a year feeding them in order to get that manure, we'd be getting no benefit from them outside of their companionship (which is a lot in our opinion).

One animal per acre is a traditional rule of thumb. But we've found that old rules of thumb, especially when it comes to raising animals, can be unreliable. Every situation is different.

You should supplement grass with minerals. Cows can typically get by in most places with only a salt lick. Llamas, donkeys, and horses should be provided with a mineral block to which they can have free access as they see fit. Horses should always have access to salt as well. But certain areas of the country are deficient in certain minerals. And if your pasture had been poorly managed before you bought it, then it may have deficiencies that you would need to supplement. This is exactly what your agricultural extension agent is for. He or she can help you figure out what nutritional needs, if any, aren't being met by your pasture. For our llamas and donkeys, we leave out one mineral lick and one salt lick at all times.

If you want to keep your fields healthy, three acres or more per animal is a safe bet. If you have enough pasture and only a few animals, you don't even need to rotate them and they can forage at will, never stressing the pasture. But if you do need to rotate, say if you have eight animals and ten acres, then you'll want to split your pasture into paddocks (using movable electric fencing is a good way) and rotate your animals every few months. It's ideal, if you're rotating, that your animals are never on the same piece of pasture twice in one year, so take that into consideration when you decide how many animals you're going to raise on the land you've got. If you've got a chicken tractor, it's a good idea for it to follow the cows wherever they've been last so as to spread their manure and till it into the ground. Everywhere your chickens go will end up having the best grass you've got.

You'll need hay for the winter. It's tempting to make your own, but haymaking takes a lot of expensive equipment and a whole lot of physical labor. So buying is probably your best option for at least the first several years you're getting started on a farm. Plan ahead though, as hay can become quite scarce at many times of the year and when the weather has affected the annual crop. It's best to buy early in the summer, after the first cutting. You never know if drought will set in and there won't be a second cutting.

First-aid Kit

The best thing to have in any first-aid kit is the number to your regular veterinarian. You should have the vet out immediately after you bring any animal to your farm to establish the relationship and have your animal checked out, much like you have a mechanic check out a car you're buying.

Every first-aid kit is different. It's not a bad idea to buy one that is already put together. You can find these at your local feed store or online. Trying to round up all the ingredients (gauze pads, wound dressing, hydrogen peroxide, iodine solution, animal thermometer, etc.) yourself is tough. But an animal first-aid kit isn't much different than a human one, which you can certainly use in a pinch.

Dealing with Death

Tom Dance, of Oak Creek Ranch in Fredericksburg, Texas, has become the undertaker of sorts for the animals from neighbors' farms. He's the only farmer within the immediate area that has a backhoe on his tractor. So whenever a horse or cow meets its timely, or untimely, end the neighbors call Tom to come dig the hole, push the animal in, and cover it up. When it's a particularly beloved animal, like a horse, the neighbors sometimes gather together and hold a funeral of sorts, saying some words of appreciation and prayer. There's sometimes a party afterwards too, just like a human wake.

A backhoe is the preferred method of taking care of a deceased large animal. If you don't know of someone with a backhoe, your vet probably does. They'll trailer it over for a fee and bury your animal. If you have to put an animal down, always try to have a vet do it. If the vet is unavailable and the task is left to you and any friends you know that can help, then you'll need to put a .22 caliber or larger bullet right in the forehead of the animal. Don't hold the barrel directly against the forehead. Be careful and quick about it.

For cows, the most common method over the years to dispose of their bodies after death is to burn them. You might not be able to dig a hole big enough in rocky ground to bury such a large animal. In fact, burning is really the best way to deal with the body of any animal as it keeps other animals from disturbing the carcass and spreading disease. It also does away with the inevitable rot and smell that might come with a shallow burial. All you need to do is pile a bunch of brush and logs on top of the animal, essentially a funeral pyre. You'd not think a big animal full of liquid would burn, but if you add enough wood, it'll burn eventually. You might also check first with your local health department to make sure this is legal in your area.

Farmer Profile

Tom and Maryneil Dance, Oak Creek Ranch, Fredericksburg, Texas

While we became involved in farming at a relatively early age, Tom and Maryneil Dance (Michael's mother and her husband) decided to take the plunge as they approached retirement. Both lived and worked in Dallas, Texas, he as an architect and she as an interior designer and furniture-store owner. They were living the hustle and bustle of city life and always rushing to that next job. Both single, they filled their time with work.

It all caught up to Tom one day and open-heart surgery was the result. As he was recovering, his doctor told him to change his workaholic lifestyle or he wouldn't be around much longer. So he sold his architecture practice and moved to the Hill Country, to Fredericksburg, Texas. He bought a piece of land, applied his building knowledge, and built a homestead. But he still had one unfinished job back in Dallas. Turns out, Maryneil Dance was working that last project too and they soon realized they had a lot more in common than wanting to flee the big city for the quiet of the country.

They quickly married and Maryneil moved to the Hill Country to join him and is now living her life-long dream, raising and riding horses, while pursuing her other passion of painting. Tom's an artist too. And they raise longhorn cattle and a couple of donkeys. They still take on a building or remodeling project here and there, working alongside each other. But they're semi-retired and living their hobby farming dream.

It wasn't all an easy trip though. When Tom began raising longhorns, he started with just a few cows. Soon, he wanted to breed them and he borrowed a bull from the friend who had gotten him started raising longhorns. The bull began to thrash about in Tom's trailer, and before he could get the truck pulled over, the bull had broken its restraints and leapt out, breaking its leg on the road. A farmer nearby spotted the trouble and came out for assistance with a gun. They put the animal down right there on the side of the road. Tom gave the farmer the animal to eat and had to pay his friend for the cost. He quickly learned that retiring on a farm is not all peace and relaxation.

Not all of his experience came from the school of hard knocks. When he first moved out to the country, Tom began having breakfast most mornings at the local diner. He'd sidle up to the counter, order some coffee, and wait for all the local farmers to come through. Inevitably, he'd strike up a conversation and he soon gained valuable advice from the local farmers who were familiar with the local land and the best practices for farming it.

When we contemplated whether we'd be able to work a farm while also working outside jobs, Tom advised us that the level of work was completely in our control. We could do as little or as much as we wanted on the farm. He told us to keep it small and the work wouldn't become overwhelming. It was the best advice we've received about farming.

Donkeys

No, they aren't stubborn. Donkeys, or long-ears, are highly intelligent, and like other highly intelligent beings, they don't just obey because they are told what to do. And if you end up being kicked by a donkey, you most likely deserved it (perhaps you called him or her an ass?). Donkeys are their own species. Male donkeys are called jacks. Female donkeys are called jennies. Mules are the product

of breeding a male donkey with a female horse. The offspring of a female donkey bred to a male horse is called a hinny. Mules and hinnies are sterile and cannot reproduce (which is a good thing for a hobby farmer). Donkeys can live to be thirty years old or more. Donkeys are known to bray, which is the hee-haw sound for which they have a reputation. Ours never did this until we put our llamas in the field with them. Now they do it mostly just to show off when a visitor arrives at the farm.

Jacks are easy to come by and are less expensive than jennies. Since we don't advocate breeding, you won't have to be as picky about the characteristics when buying a donkey. Ugly donkeys need homes too. Most people recommend that you geld males to make them less aggressive. But as long as you don't have any females around, there shouldn't be a problem, although every situation is different and jacks can be dangerous. Bring a female horse or donkey within a mile of an unfixed male and you'd better watch out and be very careful. That's another reason why you should consider staying away from breeding.

We've had two unfixed males that were raised together for eight years and never had a problem. They do wrestle quite a bit and mount each other (that's a domination thing, not a sexual one). But all that activity keeps them in shape. Unless they are really injuring each other, being aggressive to people or other animals, or if they are show animals, there should be no reason to have them fixed. Gelding donkeys is inexpensive and easy though, so if you have any concerns or have children, go ahead and have a veterinarian do it. Donkeys are strong animals that can kill a person (well, maybe not a miniature donkey), so always err on the side of caution.

Donkeys are terrific guard animals, especially if you have sheep or goats. Those ears are long for a reason—so they can hear trouble coming from miles away. We once saw a heard of gazelle at a Texas ranch being watched over by an alert donkey, standing at attention, ears scanning the distance on a hill above where the heard was resting. Our donkeys chase any stray dogs out of their field. Donkeys provide manure that is ideal for compost production. Unlike horses and cows that just defecate wherever they're standing, donkeys will create neat piles. We just drive the tractor around the field about once a month and shovel the piles into our front-end loader and pile it up to age and later add it to the garden or compost piles.

Donkeys are desert animals that originally came from Africa. On the (relatively) easy street of Hobby Farm, USA, they can easily become overweight and can founder on a field of lush grass. No one knows the real cause of founder, but it is a condition that happens to horses and donkeys, typically when they are feeding on very rich grass after a rain, in which the bone of their foot pushes into their hoof wall. This causes extreme pain and lameness, and if not caught early, can mean the end of the animal. Fat donkeys and horses are more susceptible to foundering.

Feeding

Donkeys get very wide when overweight and the crests of their necks bulge and can even fall to one side under the weight. This should be avoided at all costs because it's mostly a non-reversible condition. We made the mistake of feeding our

donkeys treats of alfalfa cubes when we first had them. We only fed them a small amount in a bucket each day. They loved it, but quickly became overweight and began to have hoof problems because of the excess protein they were receiving. As soon as we took them off feed and only allowed them to eat grass and hay in the winter, they dropped the weight and the health of their hooves improved. Under the record drought we've had this year in Virginia, we've not fed the donkeys any supplemental feed and they've never looked healthier. You may need to resort to a grazing muzzle or to keeping them in a smaller pen to keep their weight in check. If you do need to feed them grain in the winter (if you run out of hay, for instance), then buy the lowest protein mix you can find, preferably 10 percent or lower. Also, they need a mineral lick. It's the big, pinkish block available at your local feed shop. Just tell the clerk

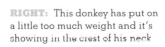
RIGHT: This donkey has put on a little too much weight and it's showing in the crest of his neck

Estimating a Donkey's Weight

To estimate an average donkey's weight using a tape measure and calculator, measure the donkey's **heart girth** in inches, which is completely around the body of the donkey a couple of inches behind his front legs. Measure the **length** from the point of the shoulder to where the tail meets the body. Measure the **height** from the top of the donkey's withers (the middle of the cross on most donkeys' back) to the ground next to the hoof. Then use this formula: Weight in pounds = heart girth (inches) x height (inches) x length (inches) / 300. This is a rough estimate and if you really need to be exact because of medication (it's okay to estimate when you're just deworming them) or some other reason, you should have a veterinarian check the weight.

you need a mineral lick for your donkeys and he or she will point you in the right direction.

The same plants that are toxic for horses are toxic for donkeys, although donkeys seem to be less susceptible to eating them in enough quantity to do much harm. You should contact your local agricultural extension agency to find out what native plants you might need to watch out for.

Maintenance

Some things you'll need to do to maintain your donkeys are to deworm them every six months, get them a tetanus shot once a year, and keep their hooves trimmed. There's some debate about whether donkeys need to be vaccinated against West Nile virus. We've never done that and our vet didn't believe it was necessary. The jury is still out on this, so you'll need to make your own decision. But donkeys don't seem to get diseases as frequently as horses and it's generally believed that they are more resistant to many of the ailments that affect horses.

To deworm a donkey, all you need to do is buy a horse deworming medication from your local feed or farm-supply store. Get the kind that you can squirt into their mouths. These also have a dial on them so you can adjust to the weight of the donkey on the applicator, insert it into the back of the donkey's mouth (go in from the side), and squeeze it in. You only need to do this about every six months.

ABOVE: Donkeys love a good roll in the dust.

Trimming their hooves is another task altogether. Donkeys are used to rocky ground that naturally keeps their hooves worn down. But most farms are all grass, so donkeys' hooves grow unimpeded. Many people have a farrier come out to trim hooves. But farriers are busy and don't like to deal with donkeys. Because donkeys don't need shoes like horses, there's no cost benefit to a farrier taking the time to come out to do it. But you should talk one into coming at least once to show you how. Or have the person you bought them from show you. It took us several lessons to learn the proper way. Neglected, donkey's hooves can grow out and curl up to resemble a genie's slippers. This is a very painful condition that requires serious veterinary help and may mean that the animal has to be put down. Hoof care is the most important task of the donkey owner. Make sure you learn it properly and keep the hooves trimmed at least every two months (every six weeks is preferable).

Donkeys can also be used for work. They can be harnessed to a cart, ridden by smaller people or children, or used to pull logs from the woods. This is something that you might explore after getting the basics of care down. But banish any ideas that you might harbor of making any money from donkeys. You might get a few bucks here and there from petting-zoo type work, but manure is the most valuable product that a donkey produces. That and protection for your other animals.

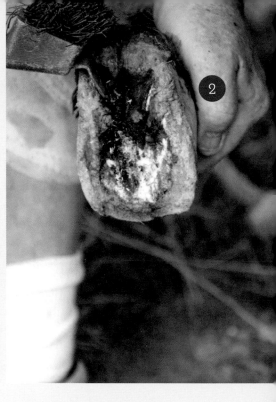

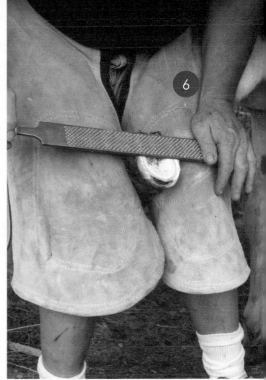

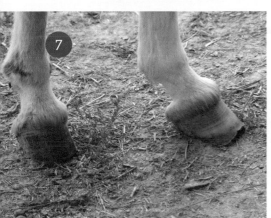

Trimming a Donkey's Hooves

1. Hoof pick, file, hoof knife and nippers.
2. Dig out manure and mud with hoof pick and brush.
3. Use hoof knife to cut away access frog (soft part of the hoof that comes to a point in the middle) material.
4. Hoof and frog should look like this when done cleaning with the knife.

5. Trim off excess hoof material all the way around the hoof. If you draw blood, back off and don't cut so much off.
6. File the hoof flat and take off any sharp material.
7. The hoof on the left has been trimmed. The hoof on the right needs it.
8. Ride your donkey into the sunset.

Do not attempt this on your own until you've had an experienced farrier come to your farm and demonstrate for you. This is the basic technique, but there are many variables and nuances that need to be demonstrated in person.

Calling a Donkey

Call, "Heeeeeee-DONK! Heeeeee-DONK!" Shake a bucket of treats in between your calls and you'll get them every time.

You should visit the American Donkey and Mule Society (ADMS) at www.love-longears.com before buying a donkey or mule and even after. They have the best resources for long-ear lovers.

Llamas and Alpacas

Our llamas always greet us first with their soft noses. They stretch their necks out, we reciprocate, and our noses come to within centimeters. We puff some air out at each other, and if we've brought the treats they are looking for, we hold them in our mouths and the llamas purse their big, soft lips and gently take them, chewing with the side-to-side motion you know from seeing camels in movies or at a zoo. Their lips are so pouty and soft; it's hard to believe they spend all day using them to gather grasses as deftly as if they were human fingers.

Llamas are very birdlike in their movements. Their slender legs and small, clawed feet seem almost too small for their big bodies and necks. But llamas' size is due mostly to their woolly fleeces, and after a shearing they reveal themselves to be quite thin and more squarely proportional. They step lightly and carefully. They appear gangly like a pubescent teenager when they are in a slow gallop, their small heads bobbing up and down every few steps. But there's nothing more graceful than a llama in a full lope. They don't run, they spring, all four feet leaving the ground together, bouncing along, heads held high, bringing to mind Tigger or Pepé Le Pew.

Llamas are more effective guard animals than our much larger and more powerful donkeys. They rarely make a sound, but when they are alarmed they emit a high call, so unique it's impossible to describe properly, but it's similar to some turkey calls. We have two llamas, and when a predator or dog enters our field, the llamas quickly come together, teaming up shoulder-to-shoulder to maximize their size. They then walk together toward the intruder and lower their heads to the ground as they walk. It's quite an intimidating show and there's not an animal yet that's ventured to test their mettle.

Llamas and alpacas are cousins. Domesticated by the Incas, they are part of the animal family of camelids. Llamas were bred mainly for use as work animals while alpacas were bred for their fiber. Alpaca wool is much finer, easier for spinners to work with, and more valuable than llama wool, in most cases. You care for each animal in pretty much the same way, taking into account that alpacas are much smaller and less able to defend themselves. If you are going to raise alpacas,

ABOVE: Ferdinand and Rico come running for their favorite apple and oat treats.

you should have a llama around to protect them. We'll only refer to llamas from here on out.

Llamas are ruminants, like cattle. But unlike cattle, they have three stomachs instead of four. Ruminants chew their cud. Chewing the cud is how ruminants digest their food. They eat it initially and it goes into their first stomach to partially digest. It's then regurgitated and they chew it again. It's the digestive juices in the first stomach that become the infamous spit that ends up on offending creatures and people. It smells about like you'd imagine partially digested food to smell. Llamas have bottom teeth and a hard palate at the top of their mouth against which they grind their food. They do have two upper pairs and one lower pair of sharp, fighting teeth. Males use these teeth to fight for the affections of females.

Llamas' reputation for spitting is probably very much a part of their evolutionary plan. We've had two gelded male llamas for five years and we've never once been spit on, but that's probably because they have no females around to arouse them. They do spit at each other occasionally, but only when they are competing for their favorite apple and oat treats. Llamas are large, imposing animals, but they rely on intimidation mostly to deter predators. They have dainty legs that, while powerful, don't carry the punch of a horse or donkey hoof. So the legend of the spit helps them keep most people at a distance.

Maintenance

There are two regular tasks that you'll need to learn to do for your llamas unless you want to pay someone to come out every six weeks to do it for you. And you should have someone show you at least once before attempting yourself. The first is to deworm the llamas.

You'll also need to learn to trim their toenails. This is easy if you can learn to grab the leg and secure it before they pull it away. You'll need to be very careful when doing the back legs as they typically will jerk their foot out of your hand and then kick you square in the behind.

RIGHT: Audrey administers the de-worming medication by injecting it into Ferdi's shoulder.

Trimming Llama Nails

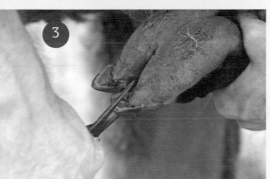
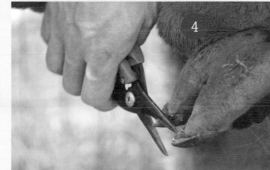

1. You can tell this llama's nails need trimming—they curl around below the pads of his foot.
2. Hold the llamas leg up and use straight clippers. Straight gardening shears work fine.
3. Trim the sides down.
4. Clip the tips. You want the nail to go straight, curve at the end and touch the ground flat with the rest of the foot.

Hand-shearing a Llama: Jonathan Sides demonstrates the proper way to hand-shear a llama. That's one skinny llama under all that hair!

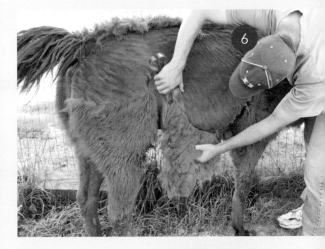

It's recommended that you have your llamas and alpacas sheared once a year. This is especially important in hot climates. Llamas are mountain animals and are easily susceptible to heat exhaustion that can lead to death. If your llama is foaming at the mouth, it's overheating and you should take steps to cool it off, including running fans on it and shearing it immediately. If you are planning to sell the fleece, you'll need to learn from someone with experience and you'll need practice. The fleece needs to be trimmed in a certain way for it to be desirable to spinners. You should have someone with experience demonstrate for you in person.

Longhorn and Grass-fed Beef Cattle

Longhorns

In Texas, there are longhorn cattle dotting the landscape everywhere you look. The local farmers call them YO, which is short for "yard ornaments." And that's why most people keep them. They look nice, don't take any real effort or cost to care for, and you can sell a couple a year at a small profit to keep your agricultural exemption. They are hardy, used to drought conditions, eat almost anything, and

BELOW: Chile the steer at Oak Creek Ranch.

very seldom have any birthing problems. Thus they are a perfect animal for the hobby farmer.

What exactly are they used for? Very few people eat them, but their hides and horns are used in clothing, furniture, and decoration. The meat mostly ends up in pet food, so you can keep them around to enjoy a long, restful life (as opposed to other cattle that are slaughtered within a couple of years) before sending them off to the butcher. If you're an animal lover, you probably have dogs and cats that need to eat. Since longhorns aren't mass-produced and are kept mostly as YO, they are probably the most humanely raised meat product around. Ironic, since they are some of the least desirable for human consumption.

Longhorns are also very resistant to disease and require a minimum amount of vaccinations and veterinary care, as long as you're not raising them in a northern climate. Longhorns do best in hot and dry areas of the country. We know farmers in Texas that have only vaccinated their longhorns once, dewormed them only every few years, and never had to call a vet for their YO; they've been raising them healthy and without incident for many years. But it's recommended that you deworm them once a year. The initial vaccinations, which you only need to do once, are quite inexpensive.

Longhorns don't need any shelter really, as long as there are trees for them to lie under. Even if you provide shelter, they probably won't use it. Animals with horns that can span eight feet don't really like to be confined indoors. They can mostly feed themselves on grass, except for the winter months. You'll need to provide them with hay during the period when there's no grass to eat. No one yet has been able to figure out exactly how much cattle need to be fed each day. So the best way to do it is to buy round bales of hay and put them in the special round bale feeders to allow the cows to eat as needed. Water, of course, is just as important, and a large water trough should be available at all times.

Longhorns generally calve without incident. The only exception to this seems to be when a larger breed bull mates with a longhorn. This happened to Tom and Maryneil Dance. A big bull from the next farm over jumped a fence and ended up mating with one of their longhorn heifers. When the time came for her to birth it, the calf was too big to pass. The cow was in obvious distress with the calf only part of the way out. This is a very common situation in most cows, but not Longhorns. After pulling on their own for awhile with no success, the Dances finally tied a rope around the calf's legs that were protruding and hooked the other end up to their all-terrain vehicle. They slowly eased it forward and the calf popped out, no worse for

wear. While this story is uncommon for longhorns, it's enough to reinforce our aversion to breeding. Having an off-farm job would make it all the more likely that you wouldn't be around to deal with an emergency like this and it could quickly turn deadly for the cow and calf.

Grass-fed Beef Cattle

If you want to add diversity to your farm operation and want to try to turn a decent profit, you might consider adding a small-scale grass-fed beef operation. It's also a good way to improve your health if you're a meat eater, as grass-fed beef is lower in fat and has more micronutrients and B vitamins than the corn-fed beef you can get in the store. But be sure that you have adequate pasture for the cows to be rotated. If you don't have enough pasture for them, you'll just throw all your profit possibilities away on feed and you'll subject your animals to greater risk of health problems.

Stick with buying steers (castrated young males) that are about two years old. Getting into breeding is more trouble than it's worth unless you decide to fully dedicate yourself to it down the road. Buy from a reputable local farmer and start with just two or three steers. Certified cattle are recommended by most farmers, but they are more expensive. Most people buying meat at the farmer's market don't care if it's certified, only that it's grass-fed and humanely raised. Angus is the best breed for a beginner and provides the meat most people are used to buying. You keep each animal for about two years and then send it off to be processed. So you'll want to buy a couple more each year to keep your yearly supply producing.

Grass-fed beef can be a sustainable part of your farm. But you must manage them carefully. You'll need several pastures that can be closed off. Usually this can be done economically with electric fencing. And you'll need to make sure you can get water to each pasture. You'll want to rotate your cows so that they are never on the same pasture twice in one year. Like everything else we suggest, start small until you know how many cows your pastures can sustainably support. One cow to three acres of pasture is a good start. Wait until they've grazed it thoroughly and move them on to the next pasture. Some people have one cow to two acres, but why go for more until you're sure about your own pasture's health?

Grass-fed beef also goes hand-in-hand with a mobile chicken coop or chicken tractor operation. Rotate the chickens around following the cows. They spread the cow manure and eat the harmful bugs (including fly larvae), all the while producing their own manure as they go. Their scratching of the dirt aerates it and works in the nutrients. If you have enough room, you can also follow the chickens with a field crop, like corn. Let the cows back in after harvest to pick through the leftovers before moving them onto grass again and you've got a healthy rotational system. But planting field crops will require that you go back and seed grass, so it's not a completely self-sustaining part of a rotational system.

Since grass-fed beef is meant for human consumption, there are government regulations that you have to follow. While some beef is labeled organic, there really is no such thing as organic beef. All cows, by law, have to be vaccinated against certain diseases, including black foot. So while your organic steak may not have any antibiotics in it (although there might even be traces of that), your beef has definitely been touched by the pharmaceutical industry. And cattle need to be dewormed. You're basically putting poison in their bodies to kill the worms. While this is allowed for an organic label, it's hardly a natural ingredient.

> *Instead of buying a horse to round up his cattle, Tom Dance bought a cheap dirt bike to round them up. One day, one of the bulls decided to challenge this strange, wheeled beast and Tom was thrown from the bike and almost seriously injured. That's when he discovered that shaking a bucket of feed cubes worked even better and was a lot less dangerous.*

Cattle must be slaughtered at a government-certified slaughterhouse. And those are getting harder and harder to find. It's unfortunate when you have to truck your cattle a couple of hundred miles to be slaughtered. So while a grass-fed operation decreases the carbon footprint of raising cattle, it doesn't eliminate it. But it's been shown that rotational beef operations can actually create a good carbon balance by returning oxygen to the air through the increase in the health of the grasses. As we've stated before, environmental purity is pretty much impossible on a working farm, but it's worth working in that direction.

Can You Make Money on Grass-fed Beef?

By now you know that there are endless variables in farming, including the market for cattle that year and the availability of hay. There will be lots of up-front costs that should be factored in over many years, like fencing and any infrastructure you lay for water delivery. As long as you stay small, grass-fed beef can be a nice supplement to your farm income. But it's also the perfect example of why you need diversity. As with all the other enterprises on your farm, it's virtually impossible to make a living by raising beef alone without becoming a factory farmer or having many hundreds of acres of healthy pasture. And keeping only cows is a good way to exhaust the natural resources on your farm. Here we provide an approximate breakdown of cost and profit potential for one cow from purchase to sale to consumer on a small-scale grass-fed beef farm:

So you can see now why farmers are constantly straining their farm system by adding more and more cattle to make a decent living. Breeding your own cattle raises your profit margin, but creates other problems and costs. One bad drought

Costs
Steer—$600 (although they can cost $1,000 or more sometimes depending on the market at the time).
Transportation—If you don't have a trailer, it'll cost at least $50 to have the steer delivered and $50 to have it shipped to the processor; $100 total.

Vaccinations and veterinary—One-time vaccination of about $100, plus once a year deworming of $20; $140 total for two years.
Supplemental hay in winter—$200 per year = $400 total.
Fly control—$25 per year = $50 total.
Processing = $400
Total Costs = $1,690
Return:
One animal will be processed into about 325 pounds of meat (more or less 25 pounds).
Selling direct-to-consumer at farmer's market (average of $6.50 per pound) = $2112.50.
Profit = $422.50 (cut that in half if you are selling it wholesale)

or one infectious disease and you're wiped out. It's the same unsustainable cycle you see in most farm operations and it's the government that usually steps in to help with farm subsidies. But this just drives prices down and the unsustainable cycle is perpetuated.

Horses

What kid hasn't grown up dreaming of owning a horse? They are the largest creatures that we can still consider as our pets. But anyone who's ever had or worked

with horses can attest that they are not pets at all. They are companions and friends. In many cases, they serve as our equals and complement our own abilities. The connection between horse and rider is powerful. So much so that horses are used very successfully in therapy. Horses will work with you. If you have hundreds of acres to tend to, a horse is the very best off-road vehicle you could have. Horses don't get stuck or have flat tires. We may need to help them survive, but only because we've decided to confine them to our farms. Given enough freedom to roam, horses thrive in the wild.

Horses are the one exception to our low-maintenance, low-cost rules of hobby farming. They can be relatively low-maintenance, but most of the heath problems we've overheard every owner bemoaning are those of their horses. And it's not cheap to feed a horse in winter or to get veterinary care for it. Unlike other animals that produce by-products (like chickens' eggs), horses rarely help offset the cost of their care. But frankly, we owe it to horses. They've been with us every step of the way since we set foot on this land (and even before), carrying our burdens, plowing our fields, and fighting our wars. Without horses, it's hard to imagine that we would have been successful at all in the New World. For that reason alone, it's worth the comparably small effort and money it takes to keep a horse happy and healthy. There are tens of thousands of horses that don't have proper homes or are being crowded out of the last remaining wild places; giving one a home is a noble endeavor.

Horses can live very happily right along with your cattle, donkeys, or llamas. They don't require any extras except perhaps some higher protein feed in the winter. As we mentioned before, they can shelter like the other animals we've discussed in three-sided sheds. But if you live in a very hot or very cold climate, it is best to have a barn with stalls. This gives you a way to get them out of the heat in the summer and possibly run fans on them to keep the flies away. And they can warm up and have easy access to food and water in the winter. They should be dewormed twice a year. It's recommended by most veterinarians and the makers of the medication that you deworm every six weeks. We feel this is excessive, but you'll want to follow your vet's advice as worm problems vary in different climates and parts of the country. You can use the same oral medication you use for the donkeys. And they need mineral and salt licks.

Fly Predators

Horses are often plagued by flies in the summer, as are cows. They swarm around their eyes and any open cuts. Traditionally, they've been controlled by using a fly spray you can buy at your local feed store. But these fly sprays are like spraying a pesticide on your animal. And many horses don't like them. Fly predators are a natural form of control. You can order them online. They are small, flying insects that feed on the larvae of flies, thus stopping their reproduction. You spread them around fresh (wet) piles of rotting manure, wet hay, or feed, where flies reproduce. Within thirty days, you should see a big difference in your fly population.

Horses, like donkeys, can founder, which is a painful condition of their hooves that can cause lameness and death. This usually happens after a rainy period that's just followed an extended dry period. The rich, sugary green grass is what makes them founder. This is another good reason to have a barn with stalls as you can control their access to the rich grass. Avoid feeding them too much grain as this can also contribute to hoof problems.

Horse hoof care and shoeing is of utmost importance. Unfortunately, this is not a task that you should do on your own. Call in a farrier. You may have a retired horse that doesn't wear shoes and you may be able to take this task on yourself eventually (it's the same process as that of a donkey), but a farrier will be needed until you are completely comfortable. Farriers are also a valuable source of all information on horses. They spend their time on other people's farms seeing and hearing what works and what doesn't. So they're worth the money in more ways that just putting shoes on your horses' hooves.

Another worry with horses is that there are many plants, trees, and weeds that are toxic to them.

RIGHT: You can tell this horse has foundered by the telltale signs of nail holes in the hoof wall where the founder boot was secured.

Maryneil Dance exercises her horse, Rio, in a round pen.

The list is quite extensive. Before getting a horse, you need to have your extension agent come to your farm and help you identify the offenders that need to be taken out. And unfortunately, herbicides are typically the only real option. You can try to mow your grass and replant heavily with healthy grasses. But get rid of them you must. There are many sad stories of horses dropping dead after grazing a small patch of the wrong food.

Unless you've rescued a horse that's injured, riding is good for you and the horse. Not only will this keep you both more in shape and healthier, it's the best way to connect with your horse and spot any problems it might be having. Riding is a social outlet for both rider and horse. Michael's mother Maryneil has a group of women friends that get together regularly to ride. For farmers who sometimes live far away from a city, riding with friends is their main social outlet and the horses generally enjoy being around each other too. Riding is therapeutic and is used more often now in programs that provide disabled children a sense of freedom and empowerment. This freedom is just as liberating for the healthy. Learning to ride is not an adventure you should undertake on your own. Failure to use the proper techniques in saddling and riding a horse can easily lead to serious injury for you or the animal. Find a stable that offers riding lessons or hire an instructor to come out to your farm and work with you and your horse.

A horse needs to exercise regularly or it risks injury. To do this, many people exercise their horse in a round pen for about thirty minutes, three times a week (although many horses get by with much less). It's also a valuable training tool and a way for you to connect with your horse without riding it. You should read a few books on proper technique and watch a few of the many videos available online. Lunging is exercising a horse in an enclosed area by encouraging it to walk, trot, and canter around you. Basically, you will alternate walking, trotting, and galloping at various speeds and different directions while teaching the horse voice commands. A round pen doesn't need to be fancy. You can just hook panels of fencing together into a circular pattern big enough for the horse to run around at a full gallop. Just make sure there are no sharp points that a horse could be injured on. Excercise the horse for several minutes one way and then switch directions in order to work both sides of the animal's muscles equally. Except for riding a horse, there's nothing more thrilling than standing in the middle of a round pen and watching a horse run full out in a circle around you. It's both frightening and exhilarating; and the horse loves to stretch its muscles in a controlled environment.

After riding or exercising, horses should be cooled down with a long walk. Feel the horses chest, between it's front legs. When it has cooled down, it's safe to hose the horse off (as long as its not freezing outside), wipe it partially dry and turn it out onto the pasture or into its stall.

Don't let the cost and time discourage you from keeping horses. You just need to keep the endeavor in the proper perspective and make sure you truly have the means to take care of a horse. You're giving the horse a good home, which it deserves, and you will connect with it on a deeper level like no other farm animal.

PART ④

Running Your Farm as a Business

CHAPTER 11

Bringing It All Together

By now we've touched on the many small enterprises you can create on your hobby farm and how they all fit together into a sustainable system. Animals turn our pastures into compost, which we use to enrich the soil for our vegetables and flowers. Bees increase pollination in our plants, increasing their yield and providing honey. Chickens turn bugs and grass into rich and tasty eggs, while mixing their manure and that of other animals into the soil to produce healthier grasses. Fungi provide mushrooms to eat while breaking down organic materials in our soil and providing nutrients to plant roots. A pond creates a wildlife habitat, while raising our water table, providing more of that lifeblood for every living thing on the farm, including the trees that we harvest selectively to heat our home. The cycle will go on as long as we will continue to nurture it.

But there's one more step that we took to make the most of our bountiful farm. All of the natural actors on the farm, including us, are helping to sustain it. But we need more than living things to keep our farm going. We need to buy tools, pay taxes and utilities, and purchase off-farm feed and supplies. And it's quite unusual for a hobby farmer only to grow all the food he or she consumes. Our off-farm jobs certainly provide for all of these things. But we want to work towards making our farm system a fully self-sustaining enterprise, so we need for it to generate income. So we took the next step and created a farm business.

A farm business should grow organically. Like building soil with compost and all-natural methods, a farm business should be built from the ground up. In the case of plants, it's unwise and short-sighted to boost them up with chemical fertilizers. This affects their hardiness in the long run, makes them susceptible to disease, and damages your soil plantation. We treat our business the same way. We haven't pumped it up prematurely and unnaturally with cash and capital outlays. We didn't grow thousands of stems of flowers until we knew we had a market for them. We broke ourselves in at local farmer's markets (beginning with the smaller ones) and listened for what the customer wanted. We are now in the largest market in the area and growing into the niches that are not being filled by others.

As we've pursued farming as a business, we've followed the advice of many before us and kept our expectations in check. Farming is full of dreamers and schemers, always chasing that elusive golden goose around one more bend to easy money. And it's fun to chase all the different ideas that sprout up when considering what to do with potentially bountiful land. But we need to be realistic. Consider how many ears of corn (at 25 cents) or fresh cut sunflower stems (at $2) it will take to buy that new tractor or any other piece of equipment. The same math needs to be applied to all farm products. Remember our estimates for small egg production? For grass-fed beef? It's only realistic to expect a few thousand dollars a year from each. But add them all up and they equal a nice, profitable business that supplements an off-farm job very well. And one day, after we've paid off our farm and the equipment, then perhaps it'll be enough to enjoy a comfortable living.

Bartering and the Farm Community

Before you dedicate yourself to farming for cash, don't forget about the incredible value that can be gained by trading with your fellow farmers. Once you get involved with a farmer's market (which we will discuss below), you'll quickly find that you'll not have to grow every single type of vegetable or farm product that you need yourself. There's a thriving economy of traded goods underlying all farmer's markets. This is very valuable in supplementing what you're doing on your farm.

Perhaps your tomatoes all had late blight this year. More than likely, not everyone has the same problem, but they might have some other problem.

And it's not just goods that are traded in the farming community, but labor. We regularly help out other farms in exchange for their help at some other time. Or we trade labor for food. There are even groups of people nowadays that get together through social networking sites to form "farm mobs." These are groups of people interested in local food and farms who will all pitch in for a day or a weekend at a farm. The labor of twenty people working together can be a huge benefit to a small farm. Imagine all the pathways that can be mulched or the barn that could be raised quickly with that many people helping. In return, the farmer usually provides a meal, drinks, and/or other farm products.

One of the true joys of farming is becoming a part of a community that helps to sustain itself. Unlike a city, where it's mostly money that's traded, a farming community becomes a larger version of all the farms it represents. While we're striving for diversity and sustainability on our farm, so are others. The web of sustainability expands to include everyone that relies on each other for goods and labor. This is a profit that cannot be

easily commoditized. And, not to be too dramatic about it, it's really the only way we'll ever get back to a well-balanced economy and a healthy environment.

The Importance of Hobby Farming Businesses

Much attention has been paid in the last few years to small farmers. While there's been an explosion in farmer's markets and the availability of local foods, it's much bemoaned that small farmers can no longer make a living; that our food system of government subsidies and distribution favors massive agri-business farms and promotes environmentally harmful farming practices. This is certainly true. But there are many small farmers that do make a good living and do practice sustainable farming. It's very hard work that requires long hours well past the typical forty-hour work week. It requires years of expertise in the sciences of growing, production, distribution, accounting, marketing, and sales. In short, it's like every other entrepreneurial enterprise.

And just like all other businesses in our current economic system (and even individuals, for that matter) the gap between the richest and the poorest has grown distressingly wide. But there is hope. It can be seen in the increased variety of foods at the grocery store, the proliferation of farmer's markets, and the inclusion of local foods on restaurant menus. A correction away from industrialized food is due and seems to be in the offing.

But it's unrealistic to imagine a time when all food will be sourced within 100 miles of your home, as is a goal of many local foodies. We've yet to come close ourselves and we have twenty-three acres in our backyard. In New York, it would be physically impossible for all the food to feed eight million people to come from the land only 100 miles to its north, south, and west. It's the same for DC, Los Angeles, or any other large city. Yes, there are some people who have done it in places with just the right population and farm mix to make it possible. But it's a small fraction of the population that lives in this ideal urban/rural mix. We do need to feed the whole country in a more sustainable manner. But without some access to markets all over the country, there won't be enough successful farmers to keep moving in a positive direction.

Even where we live, near Charlottesville, Virginia, which is a mix of small city, with a diversity of restaurants and markets, and its surrounding fertile farmland, it's difficult for a small farmer to make a living. There are only about 95,000 people in all of Albemarle County, which is the county that includes Charlottesville. At the City Market downtown on Saturdays, which is the biggest in Virginia, there are about a hundred vendors. Of those, only a handful are making a living solely from growing and selling food locally. There just aren't enough people in the area to support them fully. The way the successful farmers are making it is to obtain government grants for local farm production and alternative farm methods. They give workshops, write books, and grow big enough to sell at markets all over the region and to regional food distributors. They create value-added products like dried herbal teas or salsas and ship them all over the country. They maintain off-farm jobs. And the dirty secret of many farmer's markets is that some small farmers buy food at regional auctions and resell it at the market, thus selling the same produce you'd find in your local grocery store chain. It's the only way they can support their families. But mostly, farmer's markets are full of hobby farmers whose operations are supported by an off-farm job.

This is why hobby farming is so important. Without the eighty or so hobby farmers at our market, the diversity of the offerings there would be paltry and we would no longer be offering customers a reason to shop locally instead of at their national grocery store. Hobby farmers can focus on the specialty and niche products, like honey, flowers, mushrooms, grass-fed beef, and free-range eggs that help farmer's markets compete with grocery stores for the very limited customer base in places like Charlottesville.

How to establish your farm as a for-profit business:

- ✓ Register your farm as a business with the county.
- ✓ Incorporate your business—you should consider a limited liability corporation (LLC), if you're selling any product, like food, that could potentially harm someone or if you plan to employ outside labor.
- ✓ Maintain a separate bank account for your farm business.
- ✓ Keep organized records of your expenses and sales.
- ✓ Pay all appropriate sales taxes.
- ✓ Join local farm organizations.
- ✓ Attend farm education conferences.
- ✓ Create and maintain a website for your business.
- ✓ Sell regularly at farmer's markets and to local businesses.
- ✓ Rely on the advice of a local accountant who is familiar with the particulars of farm businesses.
- ✓ Avoid trouble with the IRS by turning a profit and paying your income taxes.

Starting a Farm Business

Hobby farming has come under close scrutiny in recent years by the IRS. Some people use a farm as a loss to lower their income tax burden they may owe through other employment. People who use farms to avoid taxes cast suspicion on all farmers. You must show a profit motive in order to be allowed to write off your farming expenses. There is no hard and fast rule, but you increase your chances of an audit by showing a farm loss for many years in a row.

The best way to avoid trouble is honesty. Don't write off your expenses until you have truly dedicated yourself to using your farm as a business. And if you are a hobby farmer that has no desire to make a profit and only use the land for your own enjoyment, then stay out of trouble with the government and just pay your taxes as you would if you were living in a suburb.

If you want to run your farm as a business, you can show a loss for several years (or longer) without running afoul of the IRS. Farming, after all, is not known for its profitability. And if you are striving to establish a reputable farm business, then you should not hesitate to use the legal tax code to the fullest benefit of your operation, especially if you're struggling to break even.

Farmer Profile

Parks Family Broadhead Mountain Farm, Cismont, Virginia

After working hard to find and build their dream house on Broadhead Mountain in Cismont, Virginia, Wally and Susan Parks soon set out to produce as much of their own food as they could to feed their three girls. They built a farm pond and stocked it with fish, established a chicken coop that produces about a dozen eggs a day, and planted a garden full of vegetables and a small grove

of apple and peach trees. They soon realized they had wild wine berries on their property that they collect in early summer. And Wally, an avid hunter, harvests a deer or two a year that he makes into homemade sausage and steaks.

Initially, the gardening was left to Wally, who enjoyed producing hundreds of pounds of tomatoes, peppers, potatoes, and beans. He cooked, preserved, and canned a basement full of produce each year and he still had buckets left over. He began giving them away to neighbors. One day, a neighbor suggested that he might try selling some of his vegetables at a local weekday farmer's market nearby. So Wally and Susan harvested all they could and Susan hauled it all in her truck for her first market. When she returned with a fat wad of cash in her pocket, Wally's eyes lit up and they've rarely missed a market day since.

A freak accident almost ended their hobby farming experience. Wally was busy working around the farm, navigating the hilly terrain in an electric all-terrain utility vehicle. It was a windy day. A gust of wind caught the top of a tree that must have been over a hundred years old. And in a remarkable bit of bad luck, Wally soon found out that a tree that falls in the woods on top of you indeed doesn't make a noise, because it knocks you out cold. Wally was severely injured and spent weeks in the hospital. His family and neighbors rallied around him and all pitched in to tend his beloved garden. Wally has made a full recovery and now has an even fuller appreciation of the bounty that his farm and family provide him.

Wally supports their hobby farming lifestyle and his two older daughters' college education by running a branch of his family's finance business. Susan now takes a more active role in the gardening and attends two or three farmer's markets a week in the summer with their daughter Audrey. The Parks' farm stand is rarely without a group of avidly appreciative customers lined up to buy the fresh produce. Wally and Susan provide some of the only truly organic ripe harvest vegetables found at the market, and their customers can't get enough. Audrey, who's eleven, pitches in and earns her allowance from a portion of the sales at the market. She's a bright addition and the customers love to interact with her. Audrey proves that the cuteness factor cannot be underestimated for increasing sales. Wally joins them at the markets when he's not working at the office or farm.

The Parks family is a perfect example of the joys that a family can experience hobby farming. They spend much quality time together sharing in the nurturing of their farm and eating all-natural, healthy food produced on their own land. Audrey learns much about gardening and running a small business, while experiencing the fruits of labor. If you can imagine a recipe for creating a natural and healthy family environment, the Parks family at Broadhead Mountain Farm has written it.

Bouquets

Ready-to-go OR Made to Order

Garden Flags

| 禾 | 土 | 米 |

$15.00 a Set

Celosia   Zinnias   Cleome

Cosmos   Amelia

CHAPTER 12

The Farmer's Market

Selling your goods at a farmer's market can be either the most rewarding experience of your life or the most demoralizing. We put a good deal of ourselves into growing and producing the things we bring to the market to sell. There's no better feeling than having a very successful day, selling almost everything we have, and reveling in the shower of customer praise. We usually head straight to lunch on these days and celebrate with a feast and several beers. The farmer's market is where every farm business should begin. We're able to get instant feedback on our products and pricing. We build a personal relationship with customers at the market and develop a farm identity. Farmer's markets are also social outlets. Some of our very best friends are the fellow farmers we've met at the market.

ABOVE: Our market is managed well, by spacing products in complimentary ways. On our row, we have flowers, vegetables, jewelry, grass-fed beef, and pottery.

But sometimes there are days when we make barely enough at market to pay our costs; everyone who's sold at a farmer's market has had days like this. People turn their noses up at our produce, complain about the high price of flowers, grill us like prosecutors about our growing process, and touch everything without buying. At the end of the day, we pack up most of what we came with and go home, tails between our legs. Because our flowers are so perishable, we give them to neighbors or use them to trade with other farmers. After an unsuccessful market day, many of our flowers and produce ultimately end up in the compost pile.

But remember Tom Martin from Poindexter Farm's advice when you have a bad day. Think of the farmer's market season as you would a baseball season. It's a long season and there are a lot of games. Every team has its winning streaks and its heartbreaking losses. Don't take the bad days personally. You may not be doing anything wrong at all. But always question your product mix and pricing. Be flexible and respond to what the customers are telling you.

Customer service is a huge part of the market. Always try to go the extra mile (if it doesn't cost you too much) to make the sale. You don't have to bargain on every sale, but don't be afraid to do it once in awhile. You have to be "on" at the market, always showing the customer that you're enjoying what you do. If you don't enjoy it, why are you at the market at all? Remember that the customer doesn't owe you anything. Just because you decided to put all the time, money, and effort into growing and selling at the market, doesn't mean you necessarily deserve to sell anything. If you are not offering the products that customers want to buy, then you need to change your approach. Don't take it out on the customers or you'll lose them before you ever have a chance to turn them into regulars.

Customers need and want to be educated about local foods and products. We really try to make a case for them to choose our locally grown flowers instead of the store-bought. When they ask us how long our flowers last, we take a moment to explain that because we grow them all outside, where they are strengthened by wind and rain, and we cut almost all of them the day before market and store them in clean buckets in our coolers, our flowers last at least ten days and sometimes as long as two weeks. The store-bought flowers mostly come from California or Guatemala, where they've been sprayed with pesticides and preservatives and shipped long distances. The only drawback is that sometimes our flowers last so long that customers don't need them every week. But we'll take their word-of-mouth praise instead as it will help us win other customers.

Diversify! It's the one common theme of every successful farming enterprise that we've come across. If you are solely dependent on one product at market, then you are subjecting yourself to too many outside risks, including competition from other producers, damaging weather, seasonal stresses, and drought. There are a few cases of farmers who've cornered a niche that no one else has filled and are successfully filling that without having to diversify—our local mushroom grower is a good example. But that takes more luck and expertise than any beginning hobby farmer is likely to have. The successful hobby farmers we know are selling several different products so that they have something to offer from the first market in April all the way to the last market in October.

And like our entire hobby farming venture, we started small. We didn't try to get into the big market with thousands of people coming through right away. We wouldn't have been able to get a spot anyway. It took us three years to get into the Saturday market we currently attend. We started with the smaller markets in the smaller towns nearby and attended weekday markets that are less demanding until we were confident that we could produce enough quality flowers to create a loyal customer base in a big market.

What will you sell?

Before you go all out to grow or produce a product you sell at the market, you should attend the markets you hope to join as a customer. If there are already five different cut-flower growers who've established a loyal customer base, you might want to reconsider making that your core product. But you'll also find out that there are many people selling different types of products as sidelines. They may focus on vegetables, but have a few buckets of sunflowers as well. Don't be discouraged by competition, but do take it into consideration. You shouldn't view other farmers as your primary competition. Your competition is the big multinational food and flower conglomerates and national grocery store chains.

Grow first for yourself to find the niche that most appeals to you. Then pursue something no one else is offering. It doesn't have to be a completely different venture. There were already a few cut flower growers at the market we joined this year. One of them has been selling at the same market for twenty years and has a devoted clientele that buys her high-end specialty flowers by the stem at very steep prices. We hope to stick around long enough to serve this part of the market eventually. The other general cut flower grower sells mostly typical wildflowers and serves the lower-price end of the market with $3, $5, and $8 bunches.

To differentiate ourselves, we focused on growing the highest quality mix of both specialty flowers and wildflowers. We focus on quality and cut the night before, or at most two days before, and store the flowers in coolers. Our customers tell us they last longer than any others at the market. Also, no one else at the market was doing any flower design. So we display many of our flowers in various bouquets of our own creation at different price points, from high-end to low-end. We also offer the customer the option of paying $2 more to take the vase with the bouquet, which many customers do. In this way, we're offering something no one else is—hardy, beautiful flowers in original designs that are ready to enjoy or give as a gift without the need to even find something in which to display them. And we offer original hand-made beaded jewelry and crafts, which certainly can't be copied by anyone else.

It's quality that is the best weapon of the hobby farmer. Because you aren't dealing in large volume, you can be discerning in what you offer and only bring the best of the best. Our neighbors at the market, Broadhead Mountain Farm, mostly sell tomatoes. Most people advise against attempting to compete with the very large growers of tomatoes, as they are the most abundant product at any market. The large growers can offer them at very low prices because they are dealing in volume, just like the grocery store. But Broadhead Mountain Farm sells out of their tomatoes every single week. Why? Because they grow them all outdoors (most commercial growers use greenhouses) and they don't pick them until they are exactly ripe. They are the reddest and tastiest at the market. They are a completely different fruit altogether than the hard tomatoes with no juice that you get at the

Examples of different product mixes:

- Poindexter Farms—Begins in early spring selling early-bearing crops like asparagus, blueberries, blackberries, and jam. They soon supplement this with eggs. And their grass-fed beef, since it must be frozen, can be sold from beginning to end. By the end of the market season, after they've sold all their beef, they finish up selling just eggs.
- Broadhead Mountain Farms—Begins by selling baked goods (like scones and muffins), greens, and wild wine berries. Then they transition away from the baked goods to traditional vegetables like tomatoes, peppers, beans, eggplant, and potatoes. They have eggs all summer too. Toward the end of the summer, when tomatoes may be past their prime, they will supplement again with baked goods.
- Ted's Last Stand (our own farm)—We begin with early bulb flowers like tulips and daffodils, potted herbs, and eggs. We also offer non-perishables like beaded jewelry, garden crafts, and farm photography to supplement the lack of early crops (all of these can be produced during the winter down time). We then sell berries into June. We go to full production of flowers from May through October, but we're constantly adding other products along the way, like honey and a small crop of peaches. By the end of the market year, the jewelry and other non-perishables again become a larger part of our sales.

larger growers' stands. The larger growers don't have the luxury of waiting until each tomato is perfectly ripe before picking. They just have to fill all their buckets, both for the market and the wholesale customers they sell to.

Our best day at market was the day we had the most diversity to offer: a full selection of cut flowers, two buckets of peaches, a new honey crop in twelve-ounce and sixteen-ounce containers, beaded jewelry, garden flags, and photo cards (of our animals and farm). We sold a good amount of everything and had a record day.

Market Management

Every farmer's market is managed differently. Some are very informal and it's a first-come, first-serve system. If this is the case, you definitely want to arrive early to get the best spot. But the best-managed markets have a dedicated market manager that is focused on offering a diverse mix of products, from vegetables, beef, and ready-to-eat foods to arts and crafts. These markets are often difficult to break into and you may be on a waiting list for a couple of years before you get in. The spots are reserved and seniority determines who gets the best spot. But you can almost always get in occasionally as an alternate vendor. Or you could do like we did and just stick with the smaller markets where you are guaranteed a spot and just wait for the time you can grow into the big market.

ABOVE: Audrey adds diversity to our offerings with beaded artwork.

Fees, Payment, and Sales Taxes

Every market has a different fee structure. Some just ask for a flat fee. Others ask for a one-time seasonal fee. But most of the bigger markets ask for a specific percentage of your sales. It's usually done on the honor system and you report your sales at the end of the market and pay your percentage.

We accept cash and checks. Bring a cash box with plenty of change. It's a terrible shame to lose a sale just because you don't have enough change for a customer. Some markets now offer ways for people to use credit cards, by purchasing tokens at the manager's table. Also, it's becoming easier and less expensive to take credit cards using smart phones. But save this method for later, once you're well established and thriving with good cash flow.

You'll also need to pay your state sales taxes. We don't charge the customer the actual sales tax at the time of sale as it would take too long to calculate each sale and we don't want to worry about changing small coins. So we just include it in the prices of our products. Most states now have online systems that you go into each month, fill out an online form, and they calculate the taxes you owe. Then you can usually just pay them right there online using your business check card. You don't need an advanced accounting program to keep your books, especially not right away. We still keep track of all our sales by writing them down and adding them manually. At most, we get fifty customers on a Saturday and it's not difficult to keep track. Once the volume gets higher, many people opt for a cash register and accounting software.

Establishing Your Farm Brand

We named our farm Ted's Last Stand after our first rooster that was killed by a fox. Customers like to ask about the name and it gives us a chance to tell the story and connect with them. It always puts a smile on their face and that can't be bad for sales. A story to go along with your name is a nice marketing tool and makes your farm easy to remember. We asked Michael's brother Paul, a graphic designer, to design a cool logo that features an image of Ted clutching a bunch of flowers. We display it on our website, business cards, and the eight-foot banner with our farm name that we hang across the front of our booth. We spent less than $100 having it made. We highly recommend an official banner or sign for your farm. Establishing yourself in the customer's mind and differentiating your farm from all the others at the market is an important tool in gaining repeat business.

Customers at the farmer's market are already inclined to shop locally and support small farmers. And they want to see and hear about your farm and your methods. Along with a memorable name, you should provide photos of your farm so customers can actually see where their produce or other farm products are grown or made. Broadhead Mountain Farm has a cork bulletin board hanging in their booth with photos of their garden and chickens. Poindexter Farm uses one big laminated blow-up of their mobile chicken coop with all the chickens grazing happily. We mount photos of our animals and farm scenes on blank cards with envelopes that we buy at a local art supply store. We sell them as cards that can be given with the flowers or jewelry that people might be buying as a gift. In this way, we make a little money on the side while customers have the chance to flip through the photo cards to get a feel for our farm.

Business cards are also a valuable and cheap marketing tool. We attach them to bouquets that we deliver and have them to hand out to people at the market. We receive many calls for flowers and jewelry by doing this.

Setup and Location

Whether a farmer's market is first-come-first-serve or an assigned spot market, location is crucial. You may have to wait a year or more to get one of the prime locations in the busiest part of the market. Until then, consistency is the best you can do. Customers remember where their preferred vendors are located. So until you get the prime foot traffic spot, developing loyal and repeat customers is the primary goal. One way to do that is to be in the same spot, week after week, so customers begin to fit you into their market routine. Because we sell flowers, which are delicate, many people save our booth as their last stop. The spot we're in now is close to all the parking, which makes it easy for customers to shop for flowers on their way out. The only drawback is that sometimes they've spent most of their money by the time they get to us. Given the choice, the high foot traffic areas are always best. But don't get too hung up on your space and how great it is in comparison to everyone else. Making the best with what you've got and keeping a positive attitude is just as important.

Unless you're lucky enough to be at a market that provides cover, you'll want to have an easy-up canopy for the market to shade your produce and customers from the summer sun. You can find these at any discount or outdoor store for under $100. Also, you'll want the plastic, fold-up market tables. They come in various sizes. We use one long eight-foot table and one shorter six-foot table and align them in a T formation. Some folks use three tables and create a horseshoe, or two along each side of the booth. It all depends on your selling style. Are you more comfortable working behind a table or are you very outgoing and like to draw customers in and

engage with them? Our T method allows us to do a little bit of both. Don't forget to bring a couple of chairs too. Even the busiest markets have their down times.

Display

Make your booth look *bountiful!* Think about how you shop at the market and the booths that attract your attention. They are clean, orderly, and have clear signage. The produce looks bountiful, stacked up and sometimes overflowing from containers. The presentation of your products is crucial. You

LEFT: Audrey from Broadhead Mountain Farm cools off after a particularly hot selling streak.

Market Supplies
- ✓ Canopy with weights to keep it from flying away on a windy day
- ✓ Tables and tablecloths
- ✓ Chairs
- ✓ Wooden boxes, bowls, and baskets
- ✓ Certified market scale, if selling produce by the pound.
- ✓ Money box with plenty of change.
- ✓ Notebook to record sales, along with typical office supplies like pens and tape
- ✓ Signage, both for your farm brand and for your products
- ✓ Bungee cords and a tarp in case you need to block the sun or wind
- ✓ Bags or other containers for the customer to take your products home
- ✓ Paper towels
- ✓ Photos of your farm displayed for the customer
- ✓ Business cards
- ✓ Cooler to keep water, drinks, and nourishment for yourselves during a long day
- ✓ A positive attitude and a smile.

should only bring your best produce and wares to the market. It only takes one half-rotten tomato to lose a customer for life.

To help achieve that bountiful look, display products on several levels as opposed to just laying them all out on a single layer on the tables. You can create at least three levels using a market table. If you have old wooden crates or wooden boxes, you can put them in front of the table and set buckets of produce on them. Along with the table level for display, you can create another layer using boxes or some kind of riser (even a board supported by bricks and covered with a cloth) on top of the tables to bring your product up closer to the customer's eye level.

Tablecloths are highly recommended. It looks much more like home and takes away that sterile feeling of the plastic tables. It also covers the supplies and empty boxes you'll need to store underneath the tables. For the same reason, consider using interesting ceramic bowls or wooden baskets for your produce. For our flowers, we have a selection of metal buckets and larger glass containers for single

stems. Then we create a selection of bouquets in various sized vases. If we're selling produce, we display it in baskets or interesting wooden boxes.

Another way of achieving a bountiful look is to group your products closer together. Don't spread them out too thin trying to fill up your tables. Only use as many tables as you can fill completely. And combine your produce or other products as they sell down. Don't allow a bowl with only a couple of peppers to sit out there looking tired and lonely. Take the bowl away and put the peppers in another bowl that could use some more bounty. As you sell down, shrink the size of your display and you'll keep that bountiful look much longer.

Proper signage is another key to doing well at market. Just think about how you walk through a busy market, looking left and right, overwhelmed by all the options, shapes, and colors. If a farmer has clear signage, your eyes automatically gravitate to it. Customers want to be educated as they aren't necessarily familiar with all the products at a farmer's market. There are heirloom vegetables and other products they may have never seen in a grocery store. Handwritten signs are good, if you have very clear handwriting. But consider printing signs and having them laminated, complete with a short description of the product and a clear price. Also, chalkboards or white boards are very helpful to announce what products you have in season and what specials you might be running.

Tout all the natural growing methods that you use. You don't need to be a certified organic farm to take advantage of growing organically. Most customers now know that being a "certified" organic farm just means that you have handled all the paperwork and navigated the government bureaucracy to obtain that certification. Certification is for large growers that want to use that label for their products in grocery stores. At a farmer's market, most customers are happy knowing that you grow everything yourself and you don't use pesticides (which we hope is true after reading this book).

Pricing

Pricing is a difficult nut to crack. On one hand, you want to keep your prices low enough to sell as much as possible, especially if you are selling perishables. But if you set your price too low, then you'll create a perceived value problem and people may not trust that they are buying a premium product.

When you're just starting out, it's best to copy what other growers are charging for similar products. You don't want to create ill will with other vendors by pricing so low that you are losing money and creating a price war with other farmers. Experiment with different pricing structures, like offering two-for-one deals or breaks on price at five or ten pieces. We sell some flowers by the stem from 50 cents to $4, some in single-variety bunches for $3 and $5, and some in full bouquets at anywhere from $7 to $20. We sell sunflowers for $2 per stem or three for $5. That way, we encourage the customer up to a higher level of purchase, while they feel like they're getting a good deal (which they are).

Going the Extra Mile

While flowers are always special, we like to add that extra touch that makes the customer feel like they've bought something unique that they can't get at another booth. Many of the other cut flower growers wrap their flowers in newspaper for the customer. While this is environmentally friendly (although the newspaper

Acceptable ways to describe your products, if indeed they do apply:
- Grown using organic methods (not certified)
- No-spray produce
- Our chickens eat all-natural feed, free of antibiotics
- Free-range, pasture-raised eggs
- Grass-fed beef free of hormones
- No pesticides or herbicides used in the growing process
- Produced using our own homemade compost
- No synthetic chemical fertilizers
- We use drip irrigation

may still end up the trash), we found that it leaves the customers with that familiar black stain on their hands from the ink. Another grower puts her flowers in plastic sleeves. While these are better than newspaper, they heat the flowers up and don't keep them wet, thus reducing their vase life. Still others sell flowers in plastic cups. We differentiate ourselves by wrapping a wet paper towel around the bottom of the stems to keep them from drying out. We then wrap the flowers in white wax paper and finish it off with a twist-tie. All of this adds about twelve cents to our cost for each bouquet. As our customers walk through the market with a beautiful bunch of flowers packaged in a classy presentation, they serve as a free advertisement.

We also offer bouquets already in vases for purchase. We buy low-cost vases at thrift stores and tell the customer that they can take it with them by adding $2 to the price of the bouquet. We don't really make any money on the vases, but no one else offers this service. To add another touch of class, we tie raffia (a natural ribbon made from the raffia palm) in a bow around the vases. With the small profit we get from the vases, we pay for these other supplies that add value to our product. So always look for ideas to spice up your presentation that will stick in the customers' minds.

When we are selling edamame (soy beans that are boiled and eaten as a tasty snack), we print out a recipe card that we include with each purchase that instructs the customer in the best way to prepare or preserve the beans. Recipes are always a nice added touch to any produce sale.

Creativity is what keeps the market interesting and fun. It's not rocket science, so don't be afraid to try anything. The worst that can happen is that you don't sell as much as you like. But we've never heard of anyone trying something creative at the market who didn't get at least some enjoyment out of doing it.

Befriend Other Farmers and the Market Manager

We're all in this together. There is a real camaraderie at the market between vendors. Again, don't think of other vendors as competition. We immediately sought out the more experienced flower grower at our market, toured her farm, and profiled her in this book. She's provided us with some valuable advice. We in turn don't grow all the same flowers that she's selling and try to be more of a complement to her flower mix.

Best Farmer's Market Techniques

- Brand your farm with a logo, good signage, and business cards.
- Create an attractive, professional, but down-home display.
- Bring only your best products and show confidence in them.
- Create a bountiful look, with multiple levels of products. Combine them as they sell down.
- Cleanliness is important, especially if you're selling foods.
- Greet the market and your customers with a positive attitude.
- Educate customers as to why your product is superior without preaching.
- Learn your repeat customers' names and use them often.
- Diversify using a product mix of both perishables and non-perishables.
- Befriend other farmers and the market manager.
- Clearly label and price your products. Customers don't like to guess.
- Don't set your prices too low. You can always lower prices, but it's very hard to raise them.
- Add that extra touch by using branded or interesting packaging for your customers' purchases.
- Do your best to accommodate special requests, but don't over-commit yourself.
- Look for the open niches in the market and try to fill them.

Trading among farmers at the market is a wonderful way to enjoy fresh, local foods as if you grew them yourself. Ideas are traded like goods and farmer's market gossip helps pass the time on the slow days. Networking among farmers is invaluable and you'll learn a great deal just shooting the breeze with your fellow vendors.

And there's no more important friend you can have at the market than the market manager. Managing a market is a mostly thankless job. Managers have to put up with a whole lot of grousing from farmers that don't like their spots or don't like the way the market is managed. There are a lot of eccentric personalities in the farm world and a market manager has to juggle them all. So be kind to them and you will in turn have a better market experience and you may eventually get that prime spot in the high traffic area.

Selling to Retail Markets and Restaurants

Let's say you've established yourself at a farmer's market or two, and you're suddenly producing more than you can sell directly to the consumer. You might want to look for more markets to sell your products. With the experience you'll gain at the farmer's market, you'll know how to sell to and educate the retailers and chefs that you'll be approaching.

You should keep in mind that chefs and retail buyers are very picky. Chefs especially aren't known for their warm and accommodating demeanors. And since it's very easy for them to order all their produce from a single wholesale source, you must give them a reason to add to their workload by dealing with a small producer.

Community Supported Agriculture (CSAs)

Community supported agriculture, or CSAs, have been a popular phenomenon of the local food movement, especially in the media. The idea is to have people buy "shares" of a farm's seasonal produce in advance. The customer then picks up a box of produce once a week with whatever is in season at that time. Typically, about $500 will get you a full share for the summer months. This money paid in advance is supposed to help the farmer buy seeds and supplies for the growing season. While this has been a good way to educate some consumers about eating locally, we think it's generally a bad idea.

First, how can farmers really attempt to compete with the overwhelming selection of a Whole Foods or organic market (with their growing local food selections) by not allowing their customers any choice of what vegetables they receive every week? Second, in what other business is it a good idea to collect and spend your customers' money before you've even created a product to give them in return? Drought, disease, injury, or any number of other calamities that might prevent you from returning the customers' investment can ruin your reputation in the local community. We've seen this happen to more than one grower. Also, what will happen is that you will create a competition between your market customers and your CSA customers. Who will get the best produce? Will it be the market customers who are paying full retail price? It's those customers that you need to win over. Or will you give the best produce to those that have already paid? If you choose this route, you'll make less at the farmer's market with your inferior produce.

For a hobby farmer especially, a CSA is a bad idea. Only the bigger growers, who have the ability to fully serve all markets, including the farmer's market, local retail, and restaurants, while still having room to grow, should attempt a CSA. And

Tips for Dealing with Retail Businesses and Restaurants

- First, be absolutely sure you can actually fill all the potential orders you may get. If you take orders you can't fill, you'll lose a customer for good.
- Only bring your very best products, free of any blemishes.
- Go to see them early in the day, before they have customers to deal with.
- Feel them out on price by suggesting the very best price you'd like to get first. If they spit out their gum in surprise, then you can begin to bargain. But you'll never get them to pay you more once you've settled on a price. So it's best to start high and work down.
- Go in on a Thursday, which is when they'll be stocking up for weekend business.

even if you do, you should figure out a way to give your customers some freedom of choice in what they pick up each week.

Websites and Social Media

Once you've established yourself, design a website and participate in other social networking websites. This is a very time-consuming project to set up and may be too much of a "time suck" for your first year or two. We waited until the winter down time last year to tackle it. But once you've got it up and running, it's a valuable way to interact with customers. We've not necessarily been able to quantify the benefit yet, but it certainly adds a level of legitimacy to our farm operation. And virtually everyone uses the web to find anything they need nowadays, so you'll want to be there and available for when someone would like to contact you. While you're at the market, you should have an e-mail sign-up sheet. Collect customer e-mail addresses and then you can periodically keep them informed about which markets you'll be at, what products you have available throughout the season, and when you're running specials.

Afterword

We've been working on our farm for almost ten years now and we still feel like we're just getting started (in a good way). There are always more flower beds to build, more wood to cut, and more chickens to tend. Every time we dream about living only on the income we generate on the farm, we break out into a cold sweat. There just doesn't seem to be a sustainable way to increase our production to the level that would pay all the costs of our mortgage, insurance, and other costs without breaking us physically.

I then remind myself of the mantra that we've adopted from the beginning—start small and don't overwhelm yourself. And although we're well past the beginning stages of our farm venture, staying small is still as important as it was in the beginning. Just this year, two well-established farms in our area were hit by economic reality: One shut down completely and the other severely cut back its production. They had both grown quickly for a number of years, establishing themselves as examples of the local food movement's rise. But they grew too big eventually and it overwhelmed them. They were one personal crisis away from ruin. Our entire economic system is undergoing a reset where industries are changing dramatically. Perhaps farming is headed in the direction of small and micro-producers, people like us who are doing a very small part. But add up the potential in all those hobby farm ventures and we might just realize a sustainable local food model.

Some people ask us how hobby farming helps solve any of the problems inherent in our current food and agricultural systems. Others believe that only farmers who rely solely on their farm income for economic survival can call themselves farmers. And the self-sufficiency movement has its purists that believe you aren't doing the earth any good unless you drop out of the consumption economy completely.

We're not in this to argue; obviously we think we're doing something good. We're not convinced that relying on monoculture production of many acres is better for the food system. And we do believe that participating in the local economy is healthier for society, and not necessarily harmful to the environment.

Hobby farming is practical. Farming isn't a career or job like any other. If you're going to farm to earn enough money to fund your life, you're bound to head down the path of exponential growth. With that growth comes large liabilities like employees, equipment and debt. Hobby farming offers a path to sustainability without the usual dangers inherent in for-profit farm enterprises.

At the market, the farmers with the biggest smiles on their faces are the Tom Martins and Wally Parks of the world. They've established a small niche for themselves and their products. They are only big enough to still do everything themselves and with the help of their families. They aren't making a full living off their farms, but they love to farm, love to sell what they produce, and love to connect with people. They're making the most of farming by turning it into one of the most positive aspects of their lives. It's not a burden, but a joy. That's what hobby farming is all about. And that's the type of farming we hope you'll pursue.

Hobby Farmers . . .

- Purchase and protect land from development.
- Dramatically increase the diversity of local products, providing more choice for local shoppers.
- Buy many of their supplies locally, create local products, and sell those products to local people. This generates several layers of local benefit, including several levels of local tax revenue.
- Protect beneficial wildlife, especially bees.
- Rebuild soil and pasture that's been depleted by traditional farming methods, removing carbon dioxide from the air and returning more oxygen.
- That raise mushrooms and fungi can help clean the soil of old toxins (bioremediation).
- Generate income and tax revenue from organic matter that can be recycled and regenerated into more income and tax revenue.
- Support local communities beyond traditional economic models by trading goods and services.
- Act like carbon offsets by funneling off-farm income from less-sustainable enterprises to more organic and sustainable on-farm enterprises.
- Offer a way for businesses to use workers more wisely, allowing a healthy work-at-home environment for the worker and the use of more freelance labor for the employer. This arrangement offers the worker another way to supplement his or her income to make non-traditional work sustainable.

BELOW: Even dogs need a place to hang their hats (and stay out of the weather.)

Appendix

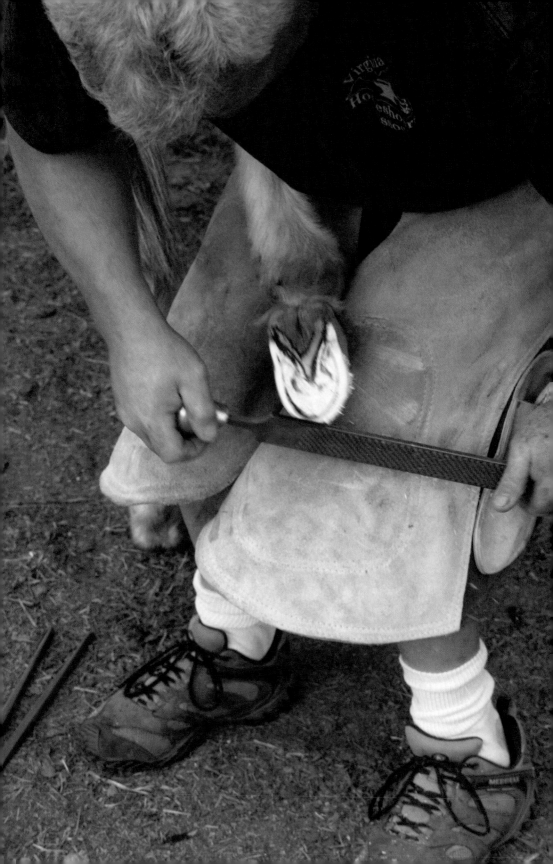

Resources

Recommended websites with searchable databases

We have found these sites to be a wealth of information on a variety of sustainable farming topics.

National Sustainable Agriculture Information Service (ATTRA)

www.attra.ncat.org

United States Department of Agriculture, Department of Food and Agriculture

www.csrees.usda.gov/Extension/index.html

Natural Resources Conservation Service (NRCS)

www.nrcs.usda.gov

Catalogs

There are many fine companies out there. This is just a small sampling of some of the catalogs we have found useful in terms of products, as well as supplemental information.

Fedco

www.fedcoseeds.com

Gardens Alive!

www.gardensalive.com

Growers Supply Company

www.growerssupply.com

Home Harvest Garden Supply

www.homeharvest.com

Johnny's Seeds

www.johnnyseeds.com

Peaceful Valley Farm and Garden Supply

www.groworganic.com

Seeds of Change

www.seedsofchange.com

Bibliography

Farm and Garden Practices and Lore

These books are just a few of our favorites. They are informative as well as thought provoking.

Coleman, Elliot. *Four Season Harvest.* White River Junction, VT: Chelsea Green, 1995.

Logsdon, Gene. *The Contrary Farmer.* White River Junction, VT: Chelsea Green, 1995.

Nearing, Helen and Scott. *The Good Life: Helen and Scott Nearing's Sixty Years of Self-Sufficient Living.* New York, NY: Schocken Books, 1989.

Part 1: Place

Philbrick, Frank and Stephen. *The Backyard Lumberjack.* North Adams, MA: Storey Publishing, 2006.

Storey, John and Martha. *Storey's Basic Country Skills.* North Adams, MA: Storey Publishing, 1999.

Salatin, Joel. *You Can Farm: The Entrepreneur's Guide to Start and Succeed in a Farming Enterprise.* Swoope, Virginia: Polyface, Inc., 1998.

Part 2: Growing Things
Soil and Compost

Lowenfels, Jeff and Lewis, Wayne. *Teeming With Microbes.* Portland, OR: Timber Press, Inc., 2010.

Pleasant, Barbara and Marin, Deborah L. *The Complete Compost Gardening Guide.* North Adams, MA: Storey Publishing, 2008.

General Gardening

Berry Hill Irrigation, www.berryhilldrip.com, Buffalo Junction, VA. Drip irrigation supplies and instructions..

Damrosch, Barbara. *The Garden Primer.* New York, NY: Workman Publishing, Inc., 2008.

Dripworks USA, www.dripworksusa.com, Willits, CA. Drip irrigation supplies and instructions.

Gillman, Jeff. *The Truth About Organic Gardening.* Portland, OR: Timber Press, Inc., 2008.

Markham, Brett L. *Mini Farming: Self-Sufficiency on a ¼ Acre.* New York, NY: Skyhorse Publishing, 2010.

Vegetable Gardening

Megyesi, Jennifer, with photography by Geoff Hansen. *The Joy of Keeping a Root Cellar.* New York, NY: Skyhorse Publishing, 2010.

Meredith, Ted Jordan. *The Complete Book of Garlic.* Portland, OR: Timber Press, Inc., 2008.

Smith, Edward C. *The Vegetable Gardener's Bible.* North Adams, MA: Storey Publishing, 2000.

Flower Farming

Byczynski, Lynn. *The Flower Farmer.* White River Junction, VT: Chelsea Green, 2008.

Berries

Yepsen, Roger. *Berries.* New York, NY: W.W. Norton & Company, 2006.

Mushrooms

Kozak, Mary Ellen and Krawczyk, Joe. *Growing Shiitake Mushrooms in a Continental Climate.* Peshtigo, WI: Field and Forest Products, Inc., 1993.

Sharondale Farm, Cismont, VA.

www.sharondalefarm.com

All the tools, spawn and knowledge you'll need to get started growing your own mushrooms.

Part 3: The Care of Living Creatures

Childs, Laura. *The Joy of Keeping Farm Animals.* New York, NY: Skyhorse Publishing, 2010.

Damerow, Gail. *Storey's Guide to Raising Chickens.* North Adams, MA: Storey Publishing, 1995.

Grandin, Temple. *Animals in Translation: Using the Mysteries of Autism to Decode Animal Behavior.* Boston, MA: Mariner Books, 2006.

Morse, Roger A. *The New Complete Guide to Beekeeping.* Woodstock, NY: The Countryman Press, 1994.

Megyesi, Jennifer, with photography by Geoff Hansen. *The Joy of Keeping Chickens*. New York, NY: Skyhorse Publishing, 2009.

American Bee Journal. www.americanbeejournal.com

Weaver, Sue. *The Donkey Companion*. North Adams, MA: Storey Publishing, 2008.

The American Donkey and Mule Society.

www.lovelongears.com

Birutta, Gale. *Storey's Guide to Raising Llamas*. North Adams, MA: Storey Publishing, 1997.

Thomas, Heather Smith. *Storey's Guide to Raising Horses*. North Adams, MA: Storey Publishing, 2000.

Acknowledgements

Much appreciation goes to our parents—Anthony and Sheila for helping us get our start, Mary for working harder on our farm than any other person, Maryneil and Tom for their sage advice and encouragement, and David for illustrating for us the joy of the place. Special thanks goes to the hardest-working people in publishing—Ann Treistman, Tony Lyons, Bill Wolfsthal, Abigail Gehring, and all the other good people at Skyhorse Publishing. We're honored that you entrusted us with this project. To all the farmers who have answered our many questions and opened their farms to our inquiring minds—Mark Jones at Sharondale Farm, the Parks family at Broadhead Mountain Farm, Tom Martin at Poindexter Farm, Eileen Stephens at Green & Gold, Rob & Krista Rahm at Forrest Green Farm, Jonathan "The Llama Whisperer" Sides, and especially our dear friends Mary and Harold Plasterer, whose combination of kind encouragement and brutally honest criticisms never fail to inform and entertain us.

Index